A FIELD GUIDE TO FOREST INSECTS AND DISEASES OF THE PRAIRIE PROVINCES

2nd Edition

Y. Hiratsuka, D.W. Langor, and P.E. Crane

SPECIAL REPORT 3

Canadian Forest Service
Northern Forestry Centre
2004

Catalogue No. Fo29-34/3-2004E-1
ISBN 0-660-19368-X
ISSN 1188-7419
First edition 1995
Second edition 2004

Distributed by:
UBC Press
c/o UNIpresses
34 Armstrong Avenue
Georgetown ON L7G 4R9
Phone: 1-877-864-8477
Fax: 1-877-864-4272

The publisher:
Canadian Forest Service
Natural Resources Canada
5320-122 Street
Edmonton AB T6H 3S5

LIBRARY AND ARCHIVES CANADA CATALOGUING IN PUBLICATION

Hiratsuka, Yasuyuki, 1933-

A field guide to forest insects and diseases of the prairie provinces
2nd ed.

(Special report, ISSN 1188-7419 ; 3)
Includes an abstract in French.
Includes bibliographical references.
ISBN 0-660-19368-X
Cat. No. Fo29-34/3-2004E-1

1. Forest insects – Prairie provinces – Handbooks, manuals, etc.
2. Trees – Diseases and pests – Prairie provinces – Handbooks, manuals, etc.
3. Conifers – Diseases and pests – Prairie provinces – Handbooks, manuals, etc.
4. Hardwoods – Diseases and pests – Prairie provinces – Handbooks, manuals, etc.
I. Langor, David William 1960- .
II. Crane, P.E. (Patricia Ellen).
III. Northern Forestry Centre (Canada)
IV. [illegible]
[illegible] hern Forestry Centre (Canada)) ; 3.

[illegible] 34.9’6’09712 C2004-980328-X

Hiratsuka, Y.; Langor, D.W.; Crane, P.E. 2004. ***A field guide to forest insects and diseases of the prairie provinces.*** **2nd Ed. Nat. Resour. Can., Can. For. Serv., North. For. Cent., Edmonton, Alberta. Spec. Rep. 3.**

ABSTRACT

The most common and important insects, diseases, and other damaging agents of forest trees in the Canadian prairie provinces are described in terms of their symptoms and signs, distribution, hosts, disease (life) cycle, and damage. These are illustrated with color photographs for easy identification. The book is divided into three major color-coded sections: one for insects and diseases of conifers, one for insects and diseases of hardwoods, and one for other damaging agents. A host index listing pest species and disorders by host species and part of host affected, a glossary of technical terms, a general reference section, and a general index listing common and scientific names of hosts and pests as well as other damaging agents are also included.

RÉSUMÉ

Les maladies, les insectes, et les autres agents nuisibles des forêts les plus répandus dans les provinces des Prairies sont décrits par leurs symptômes, leurs manifestations, leur distribution, leurs victimes, leur cycle de vie, et les dommages occasionnés. Ces phénomènes sont illustrés au moyen de photographies en couleur pour faciliter l'identification rapide. Ce livre est divisé en trois parties distingués par la couleur du papier l'une porte sur les insectes et maladies des conifères, la deuxième sur les insectes et maladies des essences feuillus, et la troisième sur les autres agents nuisibles. On y trouve également en annexe la liste des ravageurs forestiers selon l'espèce hôte et la partie de l'hôte affectée, un glossaire des termes techniques, une section de référence générale, et un index général renfermant les noms scientifiques et communs des hôtes, des ravageurs, et des autres agents nuisibles.

PREFACE TO SECOND EDITION

We were very pleased with the success of the First Edition of *A Field Guide to Insects and Diseases of the Prairie Provinces*. This book was in high demand in Canada and the USA, and was especially sought after as a resource book for forestry practitioners and for university and college courses. The first edition also garnered positive reviews as well as national and international awards for technical communication. Now that demand has exceeded supply, we are pleased to provide a second edition of this field guide. The second edition includes several revised chapters, some updated taxonomy and nomenclature, and an attractive new cover. Furthermore, the second edition is offered in both soft-cover and hard-cover formats. We would like to thank Susan Mayer for typesetting the revisions and for designing the new cover; and we thank Brenda Laishley for encouraging us to develop this second edition and for efficiently organizing all aspects of its production. Also, thanks to you, the reader and user, for your interest in this book. We hope that it will serve you well.

ACKNOWLEDGMENTS

The publication of this book was possible only through the generous support and cooperation of a large number of people. The authors are pleased to acknowledge their contributions.

This publication was encouraged and facilitated by our present program director at the Northern Forestry Centre (NoFC), S.S. Malhotra. His support has been greatly appreciated.

Portions of the manuscript were reviewed by P. Singh (Canadian Forest Service, Ottawa); J.R. Spence and P.V. Blenis (University of Alberta); A. Sproule, S.K. Ranasinghe, and H. Ono (Alberta Department of Environmental Protection); A.R. Westwood, K.R. Knowles, and Y.B. Beaubien (Manitoba Department of Natural Resources); C. Saunders (Edmonton Department of Parks and Recreation); J.P. Brandt, H.F. Cerezke, P. Chakravarty, D.W. Ip, K.I. Mallett, and D.G. Maynard (NoFC); and W.J.A. Volney (Chair of the review committee, NoFC).

Most of the photographs were taken by NoFC staff, especially by P.S. Debnam and J.H. Baker, and present and former staff of the Forest Insect and Disease Survey Unit (R.A. Blauel, H.F. Cerezke, F.J. Emond, M.G. Grandmaison, V. Hildahl, J. Petty, G.N. Still, R.C. Tidsbury). Other photographs were taken by E.A. Allen, J.C. Hopkins and G.A. Van Sickle (Pacific Forestry Centre); J.A. Muir (British Columbia Ministry of Forests); D.G. Maynard and H. Zalasky (NoFC); H. Ono (Alberta Department of Environmental Protection); P. Au (Pine Ridge Forest Tree Nursery); K.R. Knowles (Manitoba Department of Natural Resources); J.S. Scott (University of Alberta); C. Saunders (Edmonton Parks and Recreation); T.R. Reichardt (Calgary Parks and Recreation); H. Koga and J.A. Parmelee (Agriculture Canada); P.D. Syme, G.M. Howse, and R.D. Whitney (Great Lakes Forestry Centre); J. Turgeon (Forest Pest Management Institute); D.P. Ostaff (Maritimes Forestry Centre); G. Laflamme, A. Carpentier, and A. Lavallée (Laurentian Forestry Centre); and D.A. Herms (Dow Botanical Gardens, Michigan).

The following friends and colleagues at NoFC provided superb assistance with the preparation of this publication: D.D. Lee prepared the line drawings and cover art work; D.J.M. Williams helped prepare the glossary and indices; K. Jakubec helped prepare the general index; and E. Schiewe typeset the manuscript. We give special thanks to B.L. Laishley for coordinating the editing and preparation of this manuscript and for her sage advice.

This publication was prepared and printed with funds provided by the Canada–Alberta Partnership Agreement in Forestry, Canada–Saskatchewan Partnership Agreement in Forestry, and the Canada–Manitoba Partnership Agreement in Forestry.

CONTENTS

INTRODUCTION . . . 1

INSECTS AND DISEASES OF CONIFERS. . . . 3
Foliage and Buds . . . 3
Spruce budworm . . . 5
Jack pine budworm . . . 7
Other bud-feeding moths . . . 9
Larch sawfly . . . 11
Yellowheaded spruce sawfly . . . 13
Diprionid sawflies . . . 15
Spruce needleminer . . . 17
Northern lodgepole needleminer. . . . 19
Jack pine tube moth . . . 21
Larch casebearer . . . 23
Spruce spider mite . . . 25
Pine needle scale . . . 27
Spruce bud midge . . . 29
Brown felt blight . . . 31
Needle casts and other needle diseases of pine . . . 33
Elytroderma needle cast of pine . . . 37
Needle rust of hard pines . . . 39
Needle casts and other needle diseases of spruce . . . 41
Needle rusts of spruce . . . 43
Yellow witches' broom of spruce . . . 47
Needle casts and other needle diseases of fir . . . 49
Needle rusts of fir . . . 53

Roots, Stems, and Branches . . . 55
Mountain pine beetle. . . . 57
Spruce beetle. . . . 59
Other bark beetles . . . 61
Root collar weevils . . . 63
Sawyer beetles . . . 65
Northern spruce borer . . . 67
Horntails and carpenter ants . . . 69
White pine weevil . . . 71
Lodgepole terminal weevil . . . 73
Pitch blister moths . . . 75
Pine shoot borers. . . . 77
Aphids, adelgids, and scales . . . 79
Adelgid galls on spruce . . . 81
Spruce gall midge . . . 83

Lodgepole pine dwarf mistletoe . . . 85
Eastern dwarf mistletoe . . . 87
Burls of spruce and pine . . . 89
Atropellis canker of pine . . . 91
White pine blister rust . . . 93
Stalactiform blister rust of hard pines . . . 95
Comandra blister rust of hard pines . . . 97
Sweet fern blister rust of hard pines . . . 99
Western gall rust of hard pines . . . 101
Sphaeropsis (Diplodia) blight of pine and other conifers . . . 103
Scleroderris canker . . . 105
Leucostoma canker of spruce and other conifers . . . 107
Decay of conifers . . . 109
Tomentosus root rot of conifers . . . 113
Armillaria root rot . . . 115

Seedlings . . . 117
Insects attacking seedlings . . . 119
Damping-off of seedlings . . . 121
Storage molds of conifer seedlings . . . 123
Gray mold of stored conifer seedlings . . . 125

Seeds and Cones . . . 127
Coneworms and seed moths . . . 129
Cone maggots and seed wasps . . . 131
Spruce cone rust . . . 133

INSECTS AND DISEASES OF BROADLEAF TREES . . . 135
Foliage and Buds . . . 135
Forest tent caterpillar . . . 137
Large aspen tortrix . . . 139
Bruce spanworm . . . 141
Gypsy moth . . . 143
Satin moth . . . 145
Uglynest caterpillar and fall webworm . . . 147
Cankerworms . . . 149
Other common defoliating caterpillars . . . 151
Common defoliating sawflies . . . 155
Gray willow leaf beetle and American aspen beetle . . . 157
Other common leaf beetles . . . 159
Leaf skeletonizers . . . 161
Birch leafminers . . . 163
Poplar and willow leafminers . . . 165
Aphids, lace bugs, and plant bugs . . . 167
Galled and deformed leaves . . . 169
Mite galls . . . 171
Aphid galls on poplars and aspen . . . 173

Galls on willow and oak 175
Leaf spot diseases of aspen and poplar 177
Leaf and twig blight of aspen and poplar 181
Leaf rusts of aspen and poplar 183
Powdery mildew of broadleaf trees 185
Leaf and berry rusts of saskatoon and other rosaceous hosts 187
Black leaf and witches' broom of saskatoon 191
Silver leaf 193
Tar spot and black vein of willow 195

Roots, Stems, and Branches 197
Poplar borer 199
Poplar and willow borer 201
Bronze birch borer and bronze poplar borer 203
Clearwing moths 205
Smaller willowshoot sawfly 207
Scale insects 209
Ash bark beetles 211
Dutch elm disease 213
Diplodia gall and rough-bark of poplar 215
Hypoxylon canker of aspen 217
Fire blight 219
Black knot of cherry 221
Nectria and Cytospora associated with canker of broadleaf trees 223
Decay of aspen, balsam poplar, and other broadleaf trees 225

Seeds 229
Seed insects 231

OTHER DAMAGING AGENTS 233
Weather 233
Late-spring frost damage 235
Winter desiccation damage 237
Red belt 239
Needle droop of red pine 241
Wind, snow, and ice glaze damage 243
Hail damage 245
Drought damage 247
Lightning damage—circular tree mortality 249

Chemicals 251
Damage caused by herbicides, soil sterilants, and other agricultural chemicals 253
Chemical pollutant damage 255

Birds and Animals 257
Animal damage 259
Sapsucker injury 261

REFERENCES 262
GLOSSARY. 264
HOST INDEX. 272
GENERAL INDEX 284

FIGURES

1. Spruce budworm, *Choristoneura fumiferana* 4
2. Jack pine budworm, *Choristoneura pinus* 6
3. Other bud-feeding moths 8
4. Larch sawfly, *Pristiphora erichsonii*. 10
5. Yellowheaded spruce sawfly, *Pikonema alaskensis* 12
6. Diprionid sawflies 14
7. Spruce needleminer, *Taniva albolineana* 16
8. Northern lodgepole needleminer, *Coleotechnites starki*. 18
9. Jack pine tube moth, *Argyrotaenia tabulana* 20
10. Larch casebearer, *Coleophora laricella*. 22
11. Spruce spider mite, *Oligonychus ununguis* 24
12. Pine needle scale, *Chionaspis pinifoliae* 26
13. Spruce bud midge, *Rhabdophaga swainei*. 28
14. Brown felt blight caused by *Herpotrichia juniperi* 30
15. Needle casts of pine 32
16. Elytroderma needle cast of pine, caused by *Elytroderma deformans* 36
17. Needle rust of hard pines caused by *Coleosporium asterum* 38
18. A needle cast of spruce caused by *Lirula macrospora* 40
19. Needle rusts of spruce 42
20. Yellow witches' broom of spruce, caused by *Chrysomyxa arctostaphyli* 46

21. A needle cast on alpine fir caused by *Isthmiella abietis* 48
22. Needle rusts of fir 52
23. Mountain pine beetle, *Dendroctonus ponderosae* 56
24. Spruce beetle, *Dendroctonus rufipennis* 58
25. Other bark beetles 60
26. Root collar weevils 62
27. Sawyer beetles 64
28. Northern spruce borer, *Tetropium parvulum* 66
29. Horntails and carpenter ants 68
30. White pine weevil, *Pissodes strobi* 70
31. Lodgepole terminal weevil, *Pissodes terminalis* 72
32. Pitch blister moths 74
33. Pine shoot borers 76
34. Aphids, adelgids, and scales 78
35. Adelgid galls on spruce 80
36. Spruce gall midge, *Mayetiola piceae* 82
37. Lodgepole pine dwarf mistletoe, *Arceuthobium americanum* 84
38. Eastern dwarf mistletoe, *Arceuthobium pusillum* 86
39. Burls of spruce and pine 88
40. Atropellis canker of pine, caused by *Atropellis piniphila* 90
41. White pine blister rust, caused by *Cronartium ribicola* 92
42. Stalactiform blister rust of hard pines, caused by *Cronartium coleosporioides* 94
43. Comandra blister rust of hard pines, caused by *Cronartium comandrae* 96
44. Sweet fern blister rust of hard pines, caused by *Cronartium comptoniae* 98
45. Western gall rust of hard pines, caused by *Endocronartium harknessii* 100
46. Sphaeropsis blight of pine and other conifers, caused by *Sphaeropsis sapinea* 102
47. Scleroderris canker, caused by *Gremmeniella abietina* 104
48. Leucostoma canker of spruce and other conifers, caused by *Leucostoma kunzei* 106

49. Decay of conifers 108
50. Three typical types of conifer decay 111
51. Tomentosus root rot of conifers caused by *Inonotus tomentosus* 112
52. Armillaria root rot caused by *Armillaria ostoyae* 114
53. Insects attacking seedlings 118
54. Damping-off of seedlings 120
55. Storage molds of conifer seedlings 122
56. Gray mold of stored conifer seedlings, caused by *Botryotinia fuckeliana* 124
57. Coneworms and seed moths 128
58. Cone maggots and seed wasps 130
59. Spruce cone rust caused by *Chrysomyxa pirolata* 132
60. Forest tent caterpillar, *Malacosoma disstria* 136
61. Large aspen tortrix, *Choristoneura conflictana* 138
62. Bruce spanworm, *Operophtera bruceata* 140
63. Gypsy moth, *Lymantria dispar* 142
64. Satin moth, *Leucoma salicis* 144
65. Uglynest caterpillar and fall webworm 146
66. Cankerworms 148
67. Other common defoliating caterpillars 150
68. Common defoliating sawflies 154
69. Gray willow leaf beetle and American aspen beetle 156
70. Other common leaf beetles 158
71. Leaf skeletonizers 160
72. Birch leafminers 162
73. Poplar and willow leafminers 164
74. Aphids, lace bugs, and plant bugs 166
75. Galled and deformed leaves 168
76. Mite galls 170
77. Aphid galls on poplars and aspen 172

78. Galls on willow and oak 174
79. Leaf spot diseases of aspen and poplar 176
80. Leaf and twig blight of aspen and poplar 180
81. Leaf rusts of aspen and poplar 182
82. Powdery mildew of poplar and willow, caused by *Uncinula adunca* 184
83. Leaf and berry rusts of saskatoon 186
84. Black leaf and witches' broom of saskatoon, caused by *Apiosporina collinsii* 190
85. Silver leaf, caused by *Chondrostereum purpureum* 192
86. Tar spot and black vein of willow 194
87. Poplar borer, *Saperda calcarata* 198
88. Poplar and willow borer, *Cryptorhynchus lapathi* 200
89. Bronze birch borer and bronze poplar borer 202
90. Clearwing moths 204
91. Smaller willowshoot sawfly, *Euura atra* 206
92. Scale insects 208
93. Ash bark beetles 210
94. Dutch elm disease, caused by *Ophiostoma ulmi* 212
95. Diplodia gall and rough-bark of poplar, caused by *Diplodia tumefaciens* 214
96. Hypoxylon canker of trembling aspen, caused by *Entoleuca mammata* 216
97. Fire blight, caused by *Erwinia amylovora* 218
98. Black knot of cherry, caused by *Apiosporina morbosa* 220
99. Nectria and Cytospora associated with canker of broadleaf trees 222
100. Decay of aspen 224
101. Decay of balsam poplar and other broadleaf trees 226
102. Seed insects 230
103. Late-spring frost damage 234
104. Winter desiccation damage 236
105. Red belt 238
106. Needle droop of red pine 240

107. Wind and snow damage 242
108. Hail damage 244
109. Drought damage 246
110. Lightning damage—circular tree mortality 248
111. Damage caused by herbicides and soil sterilants 252
112. Chemical pollutant damage 254
113. Animal damage 258
114. Sapsucker injury 260
115. Some terms used to describe insects. 271

TABLES

1. Symptoms and signs of needle casts and other needle diseases of pine in the prairie provinces 35
2. Symptoms and signs of needle rusts of spruce in the prairie provinces 45
3. Symptoms and signs of common needle casts and other needle diseases of fir in the prairie provinces. 51
4. Descriptions of common decay fungi of coniferous trees in the prairie provinces . . . 110
5. Distribution and hosts of common defoliating caterpillars of deciduous trees in the prairie provinces 153
6. Common leaf spot diseases of aspen and poplar in the prairie provinces 179
7. Powdery mildew species common on broadleaf trees in the prairie provinces 185
8. Symptoms, signs, and alternate hosts of the four species of *Gymnosporangium* that attack saskatoon in the prairie provinces 189
9. Descriptions of common decay fungi of broadleaf trees in the prairie provinces 227

INTRODUCTION

Many biotic and abiotic agents affect the health of forest and shade trees in the prairie provinces of Canada (Alberta, Saskatchewan, and Manitoba). These agents include fungi, insects, mites, bacteria, viruses, parasitic flowering plants, climatic conditions, chemicals, mammals, and birds. Over the last 15 years several color-illustrated publications were produced at the Northern Forestry Centre to describe many of these agents (Malhotra and Blauel 1980; Hiratsuka 1987; Ives and Wong 1988; Hiratsuka and Zalasky 1993). These publications are widely recognized for their value as reference and identification guides, particularly in the office or laboratory. It is also recognized that these publications are less amenable to field use because of their number and size. Also, some of these guides are exhaustive in their coverage and include many pest species and other damaging agents that are relatively uncommon. To address these issues, we have produced a pocket-sized color-illustrated field guide to the most important and common fungi, insects, and other agents that damage forest trees in the prairie provinces.

The specific damaging agents described in this publication were selected according to the following criteria: current or potential economical importance, visually conspicuous, and well known to the public because of media exposure. Less emphasis was given to agents restricted to ornamental trees.

This publication was prepared mainly for forest management and protection personnel, pest extension personnel, park naturalists, students of post-secondary institutions (as a supplement for forest pathology/entomology or forest protection courses), high school teachers, and the general public. In consideration of this broad audience, the text is written in a simple form but is scientifically accurate and informative.

The book is divided into three major color-coded sections: insects and diseases of conifers, insects and diseases of broadleaf trees, and damage caused by other agents. The first two sections are further subdivided (not color-coded) according to the portions of trees affected. Each chapter contains basic information on symptoms and signs, distribution, hosts, disease (life) cycle, and damage for a pest species, group of species, or other damaging agent(s). Emphasis is placed on the Symptoms and Signs section, and photographs were largely chosen for their diagnostic value. References are not provided for each chapter; however, a general reference section is given at the back. No recommendations on control methods are provided. Information for small-scale control measures can be obtained from competent garden centers if the identity of the damaging agent is known. Large-scale control measures require authorization from appropriate federal and provincial agencies. These agencies can provide pertinent information about control options.

A host index listing damaging agents by host and part of host affected is provided to facilitate identification of damaging agents. A general index listing common and scientific names of hosts and pests as well as other damaging agents is also included. A glossary defines technical terms. The immature forms (larvae or nymphs) and adults of the major groups (orders) of insects that damage trees are illustrated in Figure 115 at the end of the glossary.

The following publications were used as standard sources for scientific and common names and classification:

Trees — Hosie 1979
Herbaceous plants — Moss 1959
Insects and Mites — Stoetzel 1989
Insect Classification — Borror et al. 1989
Fungi — Ginns 1986
Fungal Classification — Farr et al. 1989

In the text of each chapter the scientific name of pest organisms is placed in boldface when additional information about that species or genus is given. The first time that a scientific name of a pest organism is mentioned, the higher level classification is provided in parentheses: order and family for insects and mites, subdivision and class for Deuteromycotina (fungi), and subdivision and order for all other fungi. The synonyms of the Latin names of fungi are given only when older and more familiar names have been replaced with new ones.

The primary purpose of this guide is to facilitate identification of causes of tree damage in the field. Most of the damaging agents covered in this publication can be identified using the naked eye or a 10× hand lens. Specific identification of some fungi, however, requires examination of microscopic characters such as spore size and shape. This is possible only with a microscope fitted with an ocular micrometer.

Occasionally it may be necessary to transport samples of damaged plant tissue or damaging agents to the laboratory or to specialists for further study. Certain precautions should be taken to ensure that the samples are collected, packaged, and shipped properly to facilitate prompt diagnosis. Some general guidelines are the following:

- Collect a sufficient quantity of damaged material that illustrates both symptoms and signs.
- Collect as many specimens of feeding insects as possible, especially if they are immature, to offset high mortality during shipping and rearing. A minimum of 20 specimens is desirable.
- Provide fresh food for leaf-feeding insects, especially immature stages, to maintain them during transport.
- Adult insects (killed by freezing) or dead specimens may be preserved in 70% or more concentrated isopropyl (rubbing) alcohol and shipped in tight plastic bottles.
- Plant samples and live damaging agents should be packed so as to endure rough handling. Use a cardboard box or mailing can for packaging and shipping; never use plastic bags.
- Samples should be kept cool and shaded from direct sunlight during transport to minimize degradation and mortality.
- Ship samples as quickly as possible to their destination and inform the receiving party before the shipment is sent.
- All samples should be accompanied by detailed information regarding collection locality, date, host, parts of host affected, tree/stand/site conditions, and any other information that may be related to the problem. The name, address, and phone number of the collector should also be included.

The authors hope that you will find this field guide to be user-friendly, useful and informative. We encourage you to use it often.

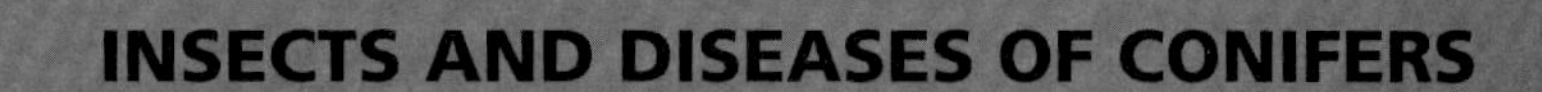

INSECTS AND DISEASES OF CONIFERS

Foliage and Buds

A
B
C
D
F
G
E
H
4

Symptoms and Signs

Larvae of the spruce budworm, ***Choristoneura fumiferana*** (Clemens) (Lepidoptera: Tortricidae), begin feeding in the buds in the spring. The first signs of damage are frass and silken webbing in buds or the last year's needles. As buds develop, larvae continue to feed in the buds and eventually the shoots. Signs of damage then appear on branch tips as silken webbing, which is spun around needles and shoots by the developing larvae for protection during feeding (D). Terminal buds that contain small larvae usually retain their caps longer than usual in June because of the adhesion of the caps by webbing. When feeding is completed in late June, tree crowns appear rust-brown because of the accumulation of partly chewed needles, dead buds, and frass held together by the silken webbing (G). Most feeding occurs in the upper crown and on terminal branch tips, where most of the new growth occurs. Normally, current-year needles and staminate flowers are the preferred food, but when populations are high, larvae may also feed on older needles, thus increasing the visual damage (H). Spruce budworm larvae have three pairs of thoracic legs and five pairs of abdominal prolegs. Young larvae are yellowish and have dark brown heads and thoracic shields (C). Mature sixth-instar larvae are 20–22 mm long and have black heads and yellowish brown bodies that are lighter in color on the sides (E). The thoracic shield is brown and separated from the head by a whitish band. There are two rows of paired whitish spots along the back.

Distribution and Hosts

This species has a boreal distribution from Newfoundland and the northeastern United States to the Yukon, Northwest Territories, and Alaska. Its principal hosts are white spruce (*Picea glauca* (Moench) Voss) and balsam fir (*Abies balsamea* (L.) Mill.).

Life Cycle

The spruce budworm has a 1-year life cycle. Adult moths appear in late June or early July. They have wingspans of about 2 cm. The fore wings are mottled dull gray or brown and the hind wings are light gray (A). Eggs are laid over a 2-week period in July. Each female lays up to 300 lime-green eggs in masses of 15–50 on the underside of needles (B). Eggs hatch in about 10 days. The hatched egg mass remains on the needles and resembles an elongate patch of white foam (similar to Fig. 2G, Jack Pine Budworm). Young larvae seek out sites under bark scales or staminate flower scars in which to spin silken hibernation shelters. Once in the hibernacula, first-instar larvae molt to the second instar and overwinter. No larval feeding occurs to this point. Second-instar larvae emerge from hibernacula from late April to mid-May and may disperse on silken threads. They first feed in old needles, unopened buds, and male flowers. Once feeding begins, larvae develop rapidly, molting four more times and reaching maturity by mid- to late June. Mature larvae usually web two or three shoots together to form tunnels wherein they pupate (F). Moths emerge from the pupae after about 10 days.

Damage

High populations of spruce budworm develop periodically and may last for several consecutive years. Outbreaks may cover up to thousands of hectares. After 3 or more consecutive years of defoliation, tree vigor is reduced, top-kill of leaders and some terminal branch shoots may occur, and stem radial growth is reduced. These effects accumulate as the infestation continues, resulting in tree mortality 5–7 years after the start of the outbreak.

Figure 1. Spruce budworm, *Choristoneura fumiferana*. A. Adult. **B.** Egg mass. **C.** Second-instar larva. **D.** Young larva and silken webbing on twig. **E.** Sixth-instar larva. **F.** Pupa. **G.** Close-up of feeding damage on spruce. **H.** Defoliated white spruce stand.

A
B
C
D
E
F
G
H
6

Symptoms and Signs

Jack pine budworm, ***Choristoneura pinus*** Freeman (Lepidoptera: Tortricidae), larvae begin mining pollen cone buds in late May to early June. The first visible signs of damage are frass pellets and silken webbing among mined cone buds. Larvae eventually extend their individual silken feeding tunnels along developing shoots (D) and begin feeding on developing foliage and seed cones. When feeding is completed in mid-July, tree crowns appear reddish brown (H) because of the accumulation of partly chewed needles and frass enmeshed in the silken webbing (F). This damage gives the tree a scorched appearance and is visible aerially. Larvae have three pairs of thoracic legs and five pairs of abdominal prolegs. Young larvae have yellowish bodies and dark brown heads. Mature larvae have light brown to black heads, and the dark brown thoracic shield has a narrow white horizontal band at the anterior end (C). The body is reddish brown with yellowish sides and has two rows of paired spots along the back. Mature larvae are 20–23 mm long.

Distribution and Hosts

This insect occurs from the maritime provinces to Alberta and in the New England and Lake States. Jack pine (*Pinus banksiana* Lamb.) is the principal host. The jack pine budworm also feeds on Scots pine (*P. sylvestris* L.), red pine (*P. resinosa* Ait.), lodgepole pine (*P. contorta* Dougl. var. *latifolia* Engelm.), and black spruce (*Picea mariana* (Mill.) B.S.P.) but does not develop to outbreak levels in natural stands of these hosts.

Life Cycle

The jack pine budworm has a 1-year life cycle. Moths appear in late July and early August. The rusty-orange fore wings are mottled with silvery bands and flecks of darker scales (A) and hind wings are gray. The wingspan is 15–28 mm. Each mated female moth may lay over 300 eggs in several egg masses on the convex surface of pine needles. Egg masses generally consist of two rows of lime-green eggs overlapping like shingles (B). The hatched egg mass remains on the needles and resembles an elongate patch of white foam (G). Newly hatched larvae seek overwintering sites under bark scales, where they spin silken hibernacula, in which they molt to the second instar before overwintering. No feeding occurs to this point. Second-instar larvae emerge from hibernacula in the spring and may disperse to other trees on silken threads before beginning to feed on pollen cone buds. Most larvae develop to the seventh instar, with the most food being consumed in the final instar. Pupation occurs in July. Pupae are reddish brown (E) and attach to the foliage with a pair of terminal hooks at the hind end. Moths emerge about 10 days after pupation.

Damage

Jack pine budworm outbreaks in natural jack pine stands are periodic, recurring about every 10 years in the prairie provinces, and usually lasting 2–4 consecutive years. Outbreaks may cover up to 2 million hectares. During outbreaks in old jack pine stands, insects may also disperse to adjacent younger stands and plantations, especially to Scots, red, jack, and lodgepole pines. The amount of damage caused by outbreaks varies considerably. Loss of foliage reduces annual increment. Defoliation for two or more consecutive years in the same stand greatly increases the risk of top-kill and mortality. Damage also seems to be influenced by the condition of the trees' roots. Root disease, deformities (often resulting from poor planting techniques), and root disturbance are associated with more severe damage. Overmaturity of stands and drought stress also affect the severity of impact by jack pine budworm.

Figure 2. Jack pine budworm, *Choristoneura pinus*. A. Adult. **B.** Egg mass. **C.** Mature larva in staminate flowers. **D.** Feeding tunnel lined with silken webbing. **E.** Pupa. **F.** Reddish brown jack pine needles caused by larval feeding. **G.** Empty egg mass. **H.** Defoliated jack pine stand.

A
B
C
D
E
F
G
8

Symptoms and Signs

Larvae of the spruce budmoth, ***Zeiraphera canadensis*** Mutuura & Freeman (Lepidoptera: Tortricidae), and the eastern blackheaded budworm, ***Acleris variana*** (Fernald) (Tortricidae), begin feeding in the buds in late spring, later moving to the new needles on the shoots. Damage caused by these two species may be confused with spruce budworm damage. Larvae of ***Z. canadensis*** and the spruce budworm secure bud caps to the growing shoot with webbing, which causes these caps to remain long after other caps are shed (C). Damage by ***Z. canadensis*** is limited to the current year's needles; however, the other two species often damage current and older foliage. Damage caused by ***Z. canadensis*** and ***A. variana*** is inconspicuous while larvae are feeding, but is easier to detect 2–3 weeks after feeding stops, when damaged needles turn reddish brown (D, G) and start to drop, exposing feeding scars (C). Spruce budworm damage is greater and more easily detected because the larvae are larger than those of the other two species. Observation of larvae greatly facilitates identification. Larvae of all three species have three pairs of thoracic legs and five pairs of abdominal prolegs. Larvae of the spruce budworm are described elsewhere (*see* Spruce Budworm). Fully grown ***Z. canadensis*** larvae are about 10 mm long and have yellowish to dark brown heads and thoracic shields (B). The body is cream to yellowish and each segment has several small dark spots, which mark the base of the setae. Fully grown ***A. variana*** larvae are about 12 mm long. The head is reddish brown to black, the thoracic shield is dark brown to black, and the body is light green (F).

Distribution and Hosts

Both species occur throughout the boreal forest. The principal host of ***Z. canadensis*** is white spruce (*Picea glauca* (Moench) Voss), but other spruces, true firs (*Abies* spp.), and Douglas-fir (*Pseudotsuga menziesii* (Mirb.) Franco) are also attacked. ***Acleris variana*** feeds mainly on white spruce, black spruce (*Picea mariana* (Mill.) B.S.P.), and balsam fir (*Abies balsamea* (L.) Mill.), but also feeds on many other conifer species.

Life Cycle

Zeiraphera canadensis has a 1-year life cycle and overwinters as eggs under bark scales at the base of the current year's shoots. Larvae emerge in late May or June and begin feeding on new buds. Larvae develop through four instars and complete feeding in July, then they drop to the ground to pupate in the litter. Moths emerge in late July and August. Moths have a wingspan of 10–13 mm. The fore wings have light and dark brown, white, and gray patches and bands; hind wings are light gray (A). Eggs are laid in groups of 1–19 under bark scales.

Acleris variana has a 1-year life cycle and overwinters as eggs. Larvae emerge in June and begin feeding on developing shoots. Larvae develop through five instars, complete feeding in July, and pupate near feeding sites. Adults emerge in August. They have a wingspan of 12–16 mm, and the fore wings are usually gray with white bands but may have black and rust patches; the hind wings are gray (E). Eggs are laid singly on the underside of needles.

Damage

Outbreaks of these species have not been reported from the prairie provinces. Visible damage by these species, especially ***Z. canadensis***, may have been interpreted as spruce budworm damage during aerial surveys. Outbreaks of these species in other areas have resulted in some tree height and radial growth loss. Feeding by ***Z. canadensis*** is also known to disfigure shoots of open-growing young trees in plantations and of ornamentals.

Figure 3. Other bud-feeding moths. A–D. *Zeiraphera canadensis*. **A.** Adult. **B.** Larva. **C.** Larval feeding scar on spruce bud. **D.** Larval feeding damage on spruce. **E–G.** *Acleris variana*. **E.** Adult. **F.** Larva. **G.** Larval feeding damage on spruce.

A
B
C
D
10
E
F
G

Symptoms and Signs

In late spring, female larch sawflies, ***Pristiphora erichsonii*** (Hartig) (Hymenoptera: Tenthredinidae), lay eggs in the new terminal twigs; these twigs then develop a characteristic curl (C), which remains for several years. The rows of egg slits and eggs in these curled twigs can be seen with the unaided eye (B). Feeding by larvae in June and early July causes most damage. From the moment they hatch, larvae feed openly on needles (E) and are thus readily visible. No webbing is produced. Larch sawfly larvae are clean feeders and leave behind few damaged, uneaten needles and little frass on the tree; therefore the foliage does not assume a reddish color, which is characteristic of spruce budworm damage. As foliage is consumed, the bark of the twigs and branches becomes increasingly visible, giving the trees a stark, red-brown appearance (G). Larvae have three pairs of thoracic legs and seven pairs of abdominal prolegs (F). Young larvae are pale green with brownish or black heads. Mature larvae are about 16 mm long and have black heads; the upper part of the body is grayish green, and the lower part is lighter green (F).

Distribution and Hosts

This species is distributed throughout the boreal forests in North America, Asia, and Europe. In North America its principal host is tamarack (*Larix laricina* (Du Roi) K. Koch), but it will also attack other native and exotic larches.

Life Cycle

Adult larch sawflies emerge from cocoons (D) in the duff over a 6- to 8-week period beginning in late May or early June. Consequently, some larvae complete development before the last adults have emerged. Most adults are females; few males are produced, and it is not known if they mate. The female is about 10 mm long, has four membranous wings, and is predominantly black with an orange band around the abdomen and orange or yellowish markings on the legs (A). Females produce eggs without mating. Eggs are laid in slits (B) cut into the elongating shoot by the female's ovipositor, a sawlike projection near the tip of the abdomen (A). The eggs are completely hidden when first laid, but after absorbing moisture they swell, showing their glistening surfaces. Larvae are gregarious and feed in colonies (E), which tend to mingle or break up as larvae grow. Larvae generally complete development in 3–4 weeks. In late June and July, larvae drop to the ground, burrow into the duff, and construct tough, leathery cocoons (D), in which the larvae overwinter. Pupation occurs in mid-spring and adults emerge in late spring. The larch sawfly normally has one generation per year; occasionally there is a partial second generation.

Damage

Outbreaks of the larch sawfly may be extensive. Such outbreaks have occurred in the prairie provinces and many other parts of Canada. Moderate to severe defoliation by the larch sawfly (G) causes a marked reduction in needle length and radial increment. Prolonged defoliation causes twig and branch mortality and may result in tree death. Trees on moist sites survive better than trees on drier sites. Weakening of trees due to repeated defoliation may also predispose them to attack by other insects and diseases.

Figure 4. **Larch sawfly, *Pristiphora erichsonii*. A.** Adult laying eggs in larch shoot. **B.** Eggs and hatching larva in shoot. **C.** Curled shoot caused by sawfly oviposition. **D.** Cocoons. **E.** Colony of young larvae. **F.** Mature larva. **G.** Tamarack defoliated by larch sawfly larvae.

A
B
C
D
E
F

Symptoms and Signs

Newly hatched larvae of the yellowheaded spruce sawfly, ***Pikonema alaskensis*** (Rohwer) (Hymenoptera: Tenthredinidae), begin feeding on new needles in early to mid-June and later move to older foliage to continue feeding. Feeding damage is usually more intense in the upper portion of the tree crown and may result in complete defoliation of the upper shoots and branches (F). After feeding is complete, the remaining partly chewed needles and needle stubs impart a brownish color and ragged appearance to the foliage (D). No webbing is produced by the larvae. Newly hatched larvae are 3–4 mm long with yellow bodies and yellowish brown heads. Mature larvae have reddish brown heads, dark green glossy bodies with light longitudinal stripes along the back and sides, three pairs of thoracic legs, and seven pairs of abdominal prolegs (B); they are 16–20 mm long. When disturbed, the larva exudes a liquid from its mouth while arching both ends of its body in a characteristic alarm reaction.

Distribution and Hosts

This species has a transcontinental distribution. Its hosts include all native and exotic species of spruce (*Picea* spp.).

Life Cycle

The adult yellowheaded spruce sawfly is stout, reddish brown, and 8–10 mm long, and has four membranous wings (A). The female has a short, sawlike projection near the tip of the abdomen that is used to cut slits in the bark for egg laying. Adults emerge from cocoons in the duff in late May and early June and begin to lay eggs when the new shoots are 2–3 cm long. Eggs are deposited singly in a slit in the bark at the base of a needle on the new growth (C); they hatch after 5–10 days. Larvae feed first on new needles and then on older foliage, and develop to maturity in 30–40 days. When mature, the larva drops to the ground and spins a tough, brown, leathery cocoon (E) around itself and overwinters in the duff. Pupation occurs the following spring and adult sawflies emerge from cocoons a few days later, completing a 1-year life cycle.

Damage

The yellowheaded spruce sawfly damages its hosts by causing repeated defoliation. Trees become susceptible to attack 3–5 years after planting and remain susceptible until they are about 8 m in height. Trees are especially vulnerable to attack when growing openly in shelterbelts, as ornamentals, in nurseries, and occasionally in young naturally regenerated open-growing forests. One year of severe defoliation causes a reduction in shoot and radial growth. Two or more years of severe defoliation results in dead branches and top-kill, and sometimes death.

Figure 5. Yellowheaded spruce sawfly, *Pikonema alaskensis*. A. Adult. **B.** Mature larva. **C.** Eggs. **D.** Close-up of sawfly defoliation on white spruce. **E.** Cocoons (adult has emerged from one). **F.** Defoliated white spruce shelterbelt.

A
B
C
D
E
F
G
H
I
J

Symptoms and Signs

Among the most common or potentially destructive diprionid sawflies (Hymenoptera: Diprionidae) in the prairie provinces are the balsam fir sawfly, ***Neodiprion abietis*** (Harris) complex; Swaine jack pine sawfly, ***N. swainei*** Middleton; ***N. maurus*** Rohwer; red pine sawfly, ***N. nanulus nanulus*** Schedl; European spruce sawfly, ***Gilpinia hercyniae*** (Hartig); and introduced pine sawfly, ***Diprion similis*** (Hartig). Larvae of all species feed openly on old foliage, usually in late spring and summer. Needles may be completely ingested, but the remaining damaged needles impart a brownish to reddish color and ragged appearance to the foliage (G). No webbing is produced. Diprionid larvae have three pairs of thoracic legs and eight pairs of abdominal prolegs. Mature ***Neodiprion*** larvae are 18–20 mm long. Larvae of ***N. abietis*** (A) have black heads and light green bodies with dark green stripes. Larvae of ***N. swainei*** (B) have brown heads, pale yellow bodies with a dark dorsal line and two black patches at the tip of the abdomen. Mature larvae of ***N. maurus*** (C) have black heads and whitish green bodies with gray subdorsal lines, black lateral spots, and a dorsal black anal patch. Larvae of ***N. n. nanulus*** (D) have dark heads and yellowish green bodies with grayish green subdorsal lines. Several other ***Neodiprion*** species in the prairie provinces appear similar to the four species described here and are difficult to identify. Mature ***G. hercyniae*** (E) larvae are 20–22 mm long, have a dark brown triangular mark on the front of the head, and a light green body with five white stripes. ***Diprion similis*** larvae (F) are 23–25 mm long and are black with many white and yellow spots.

Distribution and Hosts

Gilpinia hercyniae, which attacks spruces (*Picea* spp.), and ***D. similis***, which attacks mainly Scots pine (*Pinus sylvestris* L.) (but also other pines), were accidentally introduced from Europe. These two species and ***N. swainei***, which attacks mainly jack pine (*Pinus banksiana* Lamb.) (but also other pines), are distributed from the east coast to Manitoba. The ***N. abietis*** complex contains a group of closely related species that attack white spruce (*Picea glauca* (Moench) Voss), balsam fir (*Abies balsamea* (L.) Mill.), and Douglas-fir (*Pseudotsuga menziesii* (Mirb.) Franco), and is distributed across Canada. ***Neodiprion maurus*** infests mainly jack pine, and ***N. n. nanulus*** infests mainly red pine (*Pinus resinosa* Ait.) (both species also attack other pines); both are distributed from eastern Canada to Saskatchewan or Alberta.

Life Cycle

Neodiprion species have one generation and ***G. hercyniae*** and ***D. similis*** likely have two generations per year. ***Neodiprion abietis*** and ***N. n. nanulus*** overwinter as eggs on the needles, larvae feed on foliage from late May to July, and adults appear and oviposit in July and August. The other species overwinter as prepupae in cocoons (I) in the soil, adults appear in late May and June, and larvae feed on foliage from late June to late August. Larvae of the second generation of the two introduced species occur in August and September. In all species, eggs are laid individually in slits (H) cut into the needles with the female's sawlike ovipositor. ***Neodiprion*** adults are stout and have four membranous wings; they are light brown to black and 4–7 mm long (J). Adults of ***D. similis*** and ***G. hercyniae*** are black with yellow markings on the abdomen and thorax and are 6–8 mm long and 3 mm wide.

Damage

None of these species have caused enough defoliation to seriously damage their hosts in the prairie provinces. Localized and brief outbreaks of species in the ***N. abietis*** complex have been reported on spruce. ***Gilpinia hercyniae***, ***D. similis***, and ***N. swainei*** cause serious defoliation of conifers in eastern Canada or Europe.

Figure 6. Diprionid sawflies. A. *Neodiprion abietis* larva. **B.** *N. swainei* larva. **C.** *N. maurus* larva. **D.** *N. n. nanulus* larva. **E.** *Gilpinia hercyniae* larva. **F.** *Diprion similis* larva. **G.** *N. abietis* feeding damage on spruce. **H.** Diprionid egg scars. **I.** Cocoon. **J.** *Neodiprion* adult female.

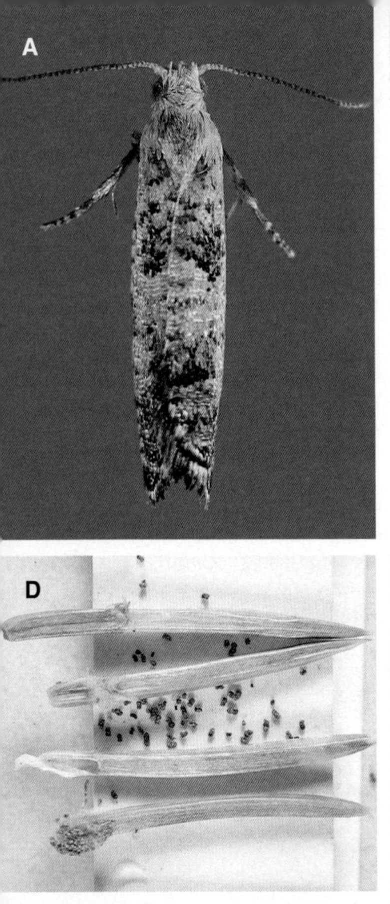
A

B

C

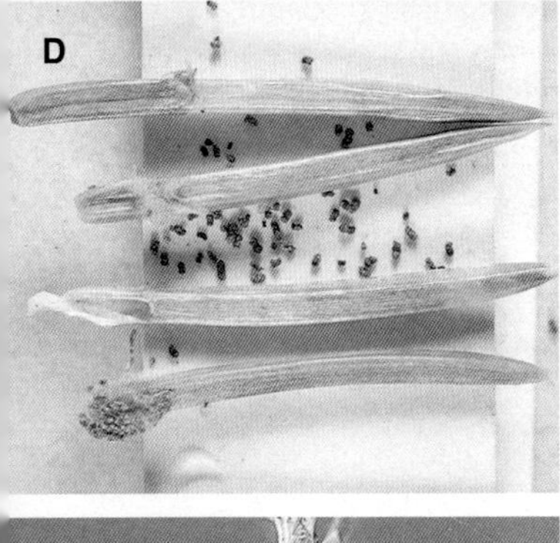
D

E

F
16

Symptoms and Signs

Damage to spruce caused by larvae of the spruce needleminer, ***Taniva albolineana*** (Kearfott) (Lepidoptera: Tortricidae), is confined to old needles. Larvae begin feeding on needles in early July. Larvae cut circular entrance holes at the base of the old needles and feed inside. Damaged needles turn pale orange (D). Larvae and brown granular frass may be seen by breaking open the needles (B, C, D). Dead needles and frass are bound together by silken webbing to form 'nests' (E, F). The nests are usually found near the base of large branches in the thickest growth. On small trees, nests are commonly located on the lowest branches next to the trunk. Nests sometimes persist on trees for a year or more. Fully grown larvae (C) are pale greenish, have dark brown heads, and are about 8 mm long.

Distribution and Hosts

This species is distributed throughout North America. The primary hosts are white spruce (*Picea glauca* (Moench) Voss), Engelmann spruce (*P. engelmannii* Parry), and Colorado blue spruce (*P. pungens* Engelm.), but many other native and exotic spruces are susceptible to attack.

Life Cycle

The spruce needleminer has one generation per year and overwinters in the larval stage. Pupation occurs in the spring and adults emerge from late May to late June. Adults are small gray and brown moths (A) with a wingspan of about 12 mm. The fore wings are marked with three irregular white bands. After mating, the female lays 3–10 pale yellow eggs along the side of a needle. Larvae hatch in about 10 days. Each larva chews a circular hole in the base of a needle, and once inside, mines upward to the tip. Usually only one larva occurs in a needle. Newly hatched larvae feed along one side of the needle, but as they grow they eat the entire needle contents. When the contents of a needle are completely consumed, the larva emerges through the entrance hole, seeks out another uninfested needle nearby, and enters that needle through its base. As they feed, groups of larvae web needles together to form nests (F). Each larva is capable of eating the contents of 10 needles. Larvae have six instars. Larvae feed until late September or October, then they spin loose cocoons inside the nest, in which they overwinter. Pupation occurs from May to early June and adult emergence occurs about 2 weeks later.

Damage

Plantations of spruce growing under adverse conditions, such as ornamentals and shelterbelts on the prairies, are particularly susceptible to serious injury. The loss of needles weakens trees by reducing growth. Nests give trees an unsightly appearance and trap airborne material such as dust and poplar and willow cotton. If spruce needleminer infestations are severe or persistent, branches become covered with a thick mat of debris that impairs the tree's vigor. Continued heavy attacks by needleminers in consecutive years, combined with damage caused by secondary organisms (fungi, insects), could cause tree death.

Figure 7. Spruce needleminer, *Taniva albolineana*. A. Adult. **B.** Needle opened to show larva in mine. **C.** Larva emerging from spruce needle. **D.** Mined spruce needles and frass. **E.** Branch of spruce damaged by spruce needleminers. **F.** Nest of mined spruce needles.

A
B
C
D
E

Symptoms and Signs

Larvae of the northern lodgepole needleminer, ***Coleotechnites starki*** (Freeman) (Lepidoptera: Gelechiidae), mine and consume the contents of needles. This causes needles to turn yellowish and then reddish brown (D). Damage to infested stands (E) is especially noticeable in even-numbered years. Close examination of affected needles reveals entrance and exit holes (C, D). These needles are hollow and contain granular frass. No webbing is produced. Early-instar larvae are lemon-yellow to light orange, with light brown thoracic and anal shields. The larvae darken progressively with each molt. Mature larvae are brownish with dark brown or black shields and heads, and are 6.5–7.5 mm long (C).

Distribution and Hosts

This species is distributed in the mountains and foothills of western Alberta as far north as Jasper and in eastern British Columbia. Populations have also been reported in the Cypress Hills of Alberta and Saskatchewan. The principal host is lodgepole pine (*Pinus contorta* Dougl. var. *latifolia* Engelm.), but ponderosa pine (*P. ponderosa* Laws.) is also attacked.

Life Cycle

This species has a 2-year life cycle and each generation overwinters twice, once in the second larval instar and once in the fourth instar. Pupation occurs in mined needles during June of the second year. Moths emerge from July to early August. Populations are synchronized, so that adults are present only in even-numbered years; however, adult emergence in odd-numbered years was observed in the Cypress Hills in the early 1980s. Adults (A) are slender with a wingspan of 12–14 mm. The fore wings are mainly light gray and usually have flecks of white, brown, dark gray, or black. Eggs are usually laid in old mines and sometimes around the needle sheaths. Newly hatched larvae mine into the tips of new needles (B) in late August and early September. After feeding, larvae molt in early October and overwinter as second instars. Development continues the next spring and most larvae finish mining the needle to its base, molt, and then transfer to a fresh needle in June or July. Fourth-instar larvae overwinter in the second year and finish mining the second needle by the following May. After molting to the fifth instar, larvae attack a third needle in late May. This needle is also completely mined by the time larval development is completed in late June.

Damage

Outbreaks of this species frequently occur in lodgepole pine stands in the foothills of Alberta (E) and may last for several years. During outbreaks defoliation may be as high as 80%. Moderate to heavy defoliation causes reduction in terminal and lateral growth. Annual increment loss of heavily defoliated stands has ranged between 20 and 75%. Tree mortality is uncommon but overmature trees may succumb during serious outbreaks.

Figure 8. Northern lodgepole needleminer, *Coleotechnites starki*. A. Adult. **B.** Needle tips mined by young larvae. **C.** Mature larva in mined needle. **D.** Close-up of mined needles. **E.** Lodgepole pine stand damaged by needleminers.

A

B

C

D

E

Symptoms and Signs

Larvae of the jack pine tube moth, ***Argyrotaenia tabulana*** Freeman (Lepidoptera: Tortricidae), begin feeding on foliage in the latter half of June. Newly hatched larvae mine the needles of the current growth. As the needle contents are eaten, the larva lines the inside of the needle with a papery, white, closely woven web to form a tube. Mined needles turn yellowish brown (C). Larvae and the silken woven tubes can be observed when mined needles are broken open. As the larvae grow, they bind other needles to the original mined needle with webbing to form a larger tube. As the needles turn color, the tube becomes conspicuous (D, E). This tube is also lined with a papery, white web and has an opening at each end. Fully grown larvae are 13–15 mm long and have yellowish brown heads and green bodies (B). The thoracic and anal shields are almost the same color as the body. A related but less abundant species, the pine tube moth, ***A. pinatubana*** (Kearfott), is similar in appearance and causes similar symptoms on eastern white pine (*Pinus strobus* L.) in southeastern Manitoba.

Distribution and Hosts

Argyrotaenia tabulana is distributed from Quebec to British Columbia in Canada and in the northern and Rocky Mountain states of the United States.The primary hosts of this species are jack pine (*Pinus banksiana* Lamb.), lodgepole pine (*P. contorta* Dougl. var. *latifolia* Engelm.), and western white pine (*P. monticola* Dougl.). There are also records of this species attacking Douglas-fir (*Pseudotsuga menziesii* (Mirb.) Franco), whitebark pine (*Pinus albicaulis* Engelm.), and western hemlock (*Tsuga heterophylla* (Raf.) Sarg.).

Life Cycle

This species overwinters in the pupal stage in loosely woven cocoons in the litter. Adults emerge in late April to early June. Adult moths have a wingspan of 13–17 mm and the fore wings are red-brown with numerous silver-gray patches (A). After mating, each female lays 2–30 eggs on the side of a needle. After eggs hatch in mid- to late June, each larva enters a needle and begins to mine its contents. When the needle is completely mined, the larva emerges, webs together several needles to form a tube, lines the tube with webbing, and continues to feed within the tube. Larvae complete feeding in late August to early September, when they drop to the ground to spin cocoons in the litter, pupate, and overwinter.

Damage

Light to moderate infestations of this species have been reported throughout the prairie provinces. In the early 1960s, one moderate to severe infestation was reported in northeastern Alberta and northwestern Saskatchewan and another near Prince Albert, Saskatchewan. Light infestations have little impact on the growth and survival of affected trees. Moderate to severe infestations significantly reduce annual radial increment and cause branch and twig mortality. Young trees may die if the infestation occurs during a period of severe drought.

Figure 9. Jack pine tube moth, *Argyrotaenia tabulana*. A. Adult. **B.** Mature larva. **C.** Tube formed from jack pine needles by larva. **D.** Close-up of damage on jack pine. **E.** Jack pine severely damaged by feeding larvae.

A
B
C

D
22

E

Symptoms and Signs

Infestations of the larch casebearer, ***Coleophora laricella*** (Hübner) (Lepidoptera: Coleophoridae), can be detected in winter by the presence of overwintering larvae in cases attached near the buds of twigs (D). These cylindrical, light brown cases consist of segments of mined larch needles and are 6–8 mm in length. Overwintering larvae begin mining larch needles in the spring as soon as new foliage appears. Foliage currently being mined is withered and brownish and has cases attached to the needles (C). Moderate to severe infestations of the larch casebearer are visible from a distance by the brownish color of the foliage (E). Mature larvae have black heads and reddish brown bodies and are about 6 mm long (B).

Distribution and Hosts

The larch casebearer was accidentally introduced into North America from Europe late in the 19th century. In Canada, the species is currently distributed as far west as eastern Manitoba, and is also present in British Columbia. The principal host is tamarack (*Larix laricina* (Du Roi) K. Koch), although other species of native and exotic larch are also attacked.

Life Cycle

The larch casebearer has one generation per year. Adults emerge, mate, and lay eggs in June. Adults are small moths with a wingspan of about 8 mm. The narrow front wings are silver or grayish brown in color, and the hind wings are fringed with hairs (A). Eggs are laid singly on the foliage. Young larvae mine the larch needles (C) until late summer, then they line segments of one or more mined needles with silk to form cases (C). The larvae carry the cases with them as they feed (B), enlarging them as necessary. When feeding, the larva fastens the fore-end of the case to a needle, which it then mines as far as possible. In the fall the larvae fasten their cases to the twigs, usually at the base of a bud, to overwinter (D). Larvae resume feeding in the spring as soon as new needles appear. Pupation occurs within the case, which is fastened with silk to a needle or twig. Moths emerge from the pupae in June to complete the life cycle.

Damage

Most of the damage is caused in the early spring when overwintered larvae resume feeding. Severely defoliated trees refoliate, and most trees are thus able to survive repeated attacks. Repeated severe defoliation causes reduction in needle length and number, reduction in radial growth, some branch and twig mortality, and occasionally tree mortality. The amount of defoliation noticed so far in Manitoba has had no discernable effect on tree growth.

Figure 10. Larch casebearer, *Coleophora laricella*. A. Adult. **B.** Larva in case. **C.** Larch needles mined by larvae (note case at arrow). **D.** Overwintering larvae in cases attached near a bud on a twig. **E.** Larch foliage damaged by larch casebearer feeding in the spring.

A
B
C
D
E
24

Symptoms and Signs

Damage to trees by the spruce spider mite, ***Oligonychus ununguis*** (Jacobi) (Acari: Tetranychidae), results from its feeding activity on old and new needles and, to some extent, on new shoots. Mites feed by inserting mouthparts into the foliar tissue and sucking out the fluids. The numerous resulting punctures give the needles a mottled appearance and cause them to dry out gradually, turn yellow to brown, and fall off prematurely. Mites also spin large amounts of silken webbing around needles and twigs (C). Dust particles, dead needles, and other debris that collect on this webbing impart a grayish color to infested branches (D) and contribute to the overall unhealthy appearance of the tree (E). The presence of silken webbing usually indicates a high population of mites. Feeding damage is first noticed in mid- to late May, and usually increases as the summer progresses. At first, damage is noticeable on the inner portion of the lower crown branches; it then spreads upward and outward in the tree crown (E). The spruce spider mite is scarcely visible to the unaided eye. To confirm a spider mite infestation one might use a hand lens to inspect branches. Another method is to hold a sheet of white paper under a branch while tapping the branch to dislodge the mites; if present, mites will be seen crawling on the paper. Larvae are oval and have three pairs of legs. At first, larvae are pinkish but turn green after feeding on foliage. Nymphs and adults have four pairs of legs and vary from dark green (B) to dark brown. Female adults are about 0.5 mm long and males are smaller.

Distribution and Hosts

The spruce spider mite has a transcontinental distribution in North America and is widely distributed in Europe. It most commonly attacks white spruce (*Picea glauca* (Moench) Voss) and blue spruce (*P. pungens* Engelm.); however, other native and exotic spruces and other conifer species are also readily attacked.

Life Cycle

Eggs overwinter under bark scales or buds or at the base of needles; they hatch in May and early June. Larvae feed by inserting their mouthparts into the needles and shoots and sucking out the sap. In about 3 days the larvae molt to a nymph stage that closely resembles the adult form. After feeding for about 6 days the nymphs molt to the adult stage (B) in June. Females may live for about 1 month and each is capable of laying 40–50 eggs (A). Eggs develop to adults in 2–3 weeks, depending upon weather. Six or more generations may occur during the summer. Hot, dry conditions appear to favor survival, and large populations may build up by late summer. Deposition of overwintering eggs begins in September and continues until a hard frost occurs.

Damage

Feeding of mites results in premature loss of foliage (E); this may kill twigs, especially in shaded areas, and may lead to death of entire branches if the infestation persists. Extremely severe infestations may kill trees, especially during periods of drought. The spruce spider mite causes the most damage on conifers planted as ornamentals, in shelterbelts, and in urban parks; in these situations, high infestations tend to persist over many years. Infestations may occasionally occur in nurseries and in natural forests, but these are usually short-lived and result in little permanent injury. Within the prairie provinces, the most intense, long-term infestations have occurred in the southern prairie and parkland zones.

Figure 11. Spruce spider mite, *Oligonychus ununguis*. A. Eggs (see arrows) on white spruce twig. **B.** Adult mite. **C.** White spruce twig severely infested with mites. **D.** Relative appearance of healthy (left) and mite-infested (right) white spruce foliage. **E.** Spruce severely infested by spider mites.

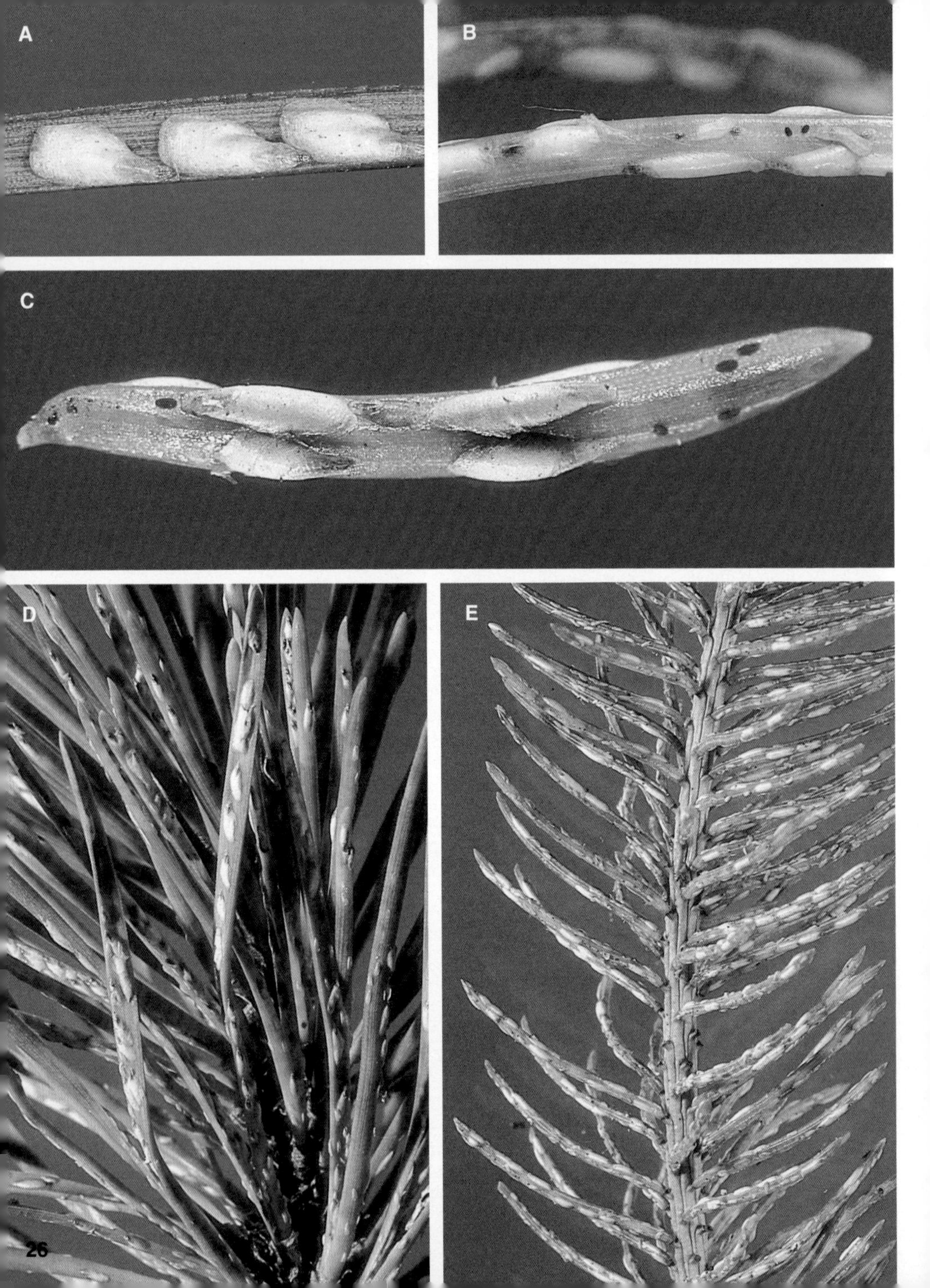
A
B
C
D
E

Symptoms and Signs

The first indication of attack by the pine needle scale, ***Chionaspis pinifoliae*** (Fitch) (Homoptera: Diaspididae), is the appearance of waxy, oval white specks or scales (about 3 mm long) on needles (D, E). When this pest becomes abundant, the needles of infested trees look white from a distance. Feeding causes yellowish green spots on the needles (D). During severe infestations the spots merge, the affected needles dry out and drop prematurely, and some or most of the foliage (except for new needles) takes on a gray, unhealthy appearance. On severely infested trees the needles are reduced to about half their normal size.

Distribution and Hosts

This species is found throughout North America and has numerous conifer hosts. In the prairie provinces this insect most commonly attacks white spruce (*Picea glauca* (Moench) Voss) and Colorado blue spruce (*P. pungens* Engelm.), but other spruces, pines (*Pinus* spp.), and Douglas-fir (*Pseudotsuga menziesii* (Mirb.) Franco) are also attacked.

Life Cycle

The pine needle scale overwinters in the egg stage beneath the protective scale secreted by the female. Most eggs hatch in mid-June. The newly hatched nymphs (crawlers) are minute, oval, and reddish (C). They wander over the needles searching for suitable feeding sites, and they may be dispersed to nearby trees by the wind. The insects remain stationary during feeding, inserting their mouthparts into the needle to feed on juices in the mesophyll. Shortly after feeding begins the nymphs flatten, turn light brown, and lose their appendages by molting. Females molt again in about 3 weeks and become sexually mature in late July or August. The mature female is wingless and about 2.5 mm long. Male scales have wings and are much smaller at 1 mm (B). Males emerge from their scale coverings in late July or August to mate, and then die. During this mating period males can frequently be seen hovering in swarms in sunny, protected areas between infested trees. After mating, each adult female secretes a waxy, white scale covering, about 6–8 mm long (A, B), which is completed in mid-August; then egg laying begins. As egg laying continues, the females gradually shrink in size and die. About 30–50 eggs fill the cavity beneath each scale, where they overwinter. There is one generation of pine needle scales per year in the prairie provinces.

Damage

Trees are usually attacked when they are young, with the most severe damage occurring in nurseries, shelterbelts, and ornamental plantings. This insect is not a serious threat to natural forests. Continuing infestations reduce the vigor and annual growth of trees, making them more susceptible to damage by other agents such as insects and diseases. Prolonged severe infestations may eventually kill twigs, branches, or small trees.

Figure 12. Pine needle scale, *Chionaspis pinifoliae*. A. Female adults. **B.** Male and female scales (males are smaller). **C.** Old scales and young crawlers (reddish color). **D.** Heavily infested pine foliage (note chlorotic spots on needles). **E.** Heavily infested spruce foliage.

A

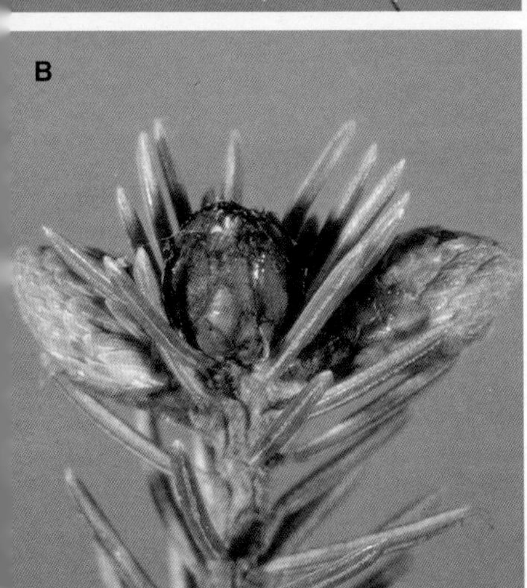
B

D

C

E

Symptoms and Signs

Feeding by the larvae of the spruce bud midge, ***Rhabdophaga swainei*** Felt (Diptera: Cecidomyiidae), kills the terminal buds of spruce leaders and branches. External symptoms of damage are not evident until the spring following the year of infestation. Infested buds are usually flatter and wider at the apex and more pinkish than noninfested buds (B). When growth continues the next year, the lateral shoots surrounding the dead bud compete for dominance, resulting in multiple leaders (E). Bud death caused by this insect is sometimes confused with that due to early spring frosts. In early spring, buds killed by spruce bud midge contain larvae or pupae, which may be seen by slicing open the dead buds (C). Fully grown larvae and pupae are orange in color and are about 2 mm long. In late spring and summer, buds killed by spruce bud midge exhibit adult emergence holes (D).

Distribution and Hosts

The spruce bud midge is distributed throughout the boreal forest. It attacks white spruce (*Picea glauca* (Moench) Voss), black spruce (*P. mariana* (Mill.) B.S.P.), and red spruce (*P. rubens* Sarg.).

Life Cycle

The spruce bud midge overwinters as late-instar larvae in cells formed in the terminal bud (C). Pupation occurs in the bud (D) in May and adults emerge in late spring. Adults are long-legged, two-winged, reddish brown flies, about 2.5 mm long (A). Females lay eggs between the needles near the tips of elongating shoots, usually one egg per shoot. After hatching, young larvae move to the tips of the elongating shoots, which they enter by burrowing into the soft base of a needle. The larvae probably feed very little until current shoots are fully developed. By late fall the larvae are nearly fully grown and overwinter. There is one generation per year.

Damage

The spruce bud midge prefers the sunny side of open growing spruce trees 2–4 m in height. Growth of the affected buds is arrested, resulting in a bushy growth form. Multiple leaders may result if the terminal bud is attacked (E), and this may lead to forked or branched trunks. Some height loss occurs, but this is usually minimal.

Figure 13. Spruce bud midge, *Rhabdophaga swainei*. A. Adult. **B.** Infested spruce bud in early spring (note rosette shape). **C.** Cross section of spruce bud to show larva. **D.** Adult emergence hole in a dead spruce bud. **E.** Black spruce with multiple leaders resulting from bud death caused by spruce bud midge.

A

B
30

C

Symptoms and Signs

The brown to black, feltlike mats of mycelium covering the lower branches of conifers are conspicuous and easily recognized (A–C). Affected needles become discolored and die.

Cause

Two fungi are known to cause brown felt blight of conifers, which is also known as snow mold. They are ***Herpotrichia juniperi*** (Duby) Petrak (= *H. nigra* R. Hartig) and ***Neopeckia coulteri*** (Peck) Sacc. (≡ *H. coulteri* (Peck) Bose) (Ascomycotina: Dothideales).

Distribution and Hosts

In the prairie region, both species of fungus occur most frequently at high elevation in the Rocky Mountains, although they are also known to exist in northern parts of the region. Hosts of ***H. juniperi*** are alpine fir (*Abies lasiocarpa* (Hook.) Nutt.), white spruce (*Picea glauca* (Moench) Voss), Engelmann spruce (*P. engelmannii* Parry), and species of juniper (*Juniperus* spp.). ***Neopeckia coulteri*** occurs only on pine. The main hosts in the prairie provinces are lodgepole pine (*Pinus contorta* Dougl. var. *latifolia* Engelm.) and limber pine (*P. flexilis* James).

Disease Cycle

Brown felt blight develops during the winter on lower branches of host trees that are covered by snow for a long time. Growth of fungi is apparently favored by the high humidity and relatively low temperature beneath the snow. The pathogens are probably disseminated by airborne ascospores or by mycelial growth from adjacent infected needles.

Damage

Although the appearance of these diseases is striking, damage is minor. Affected branches lose needles and often die, usually with no significant effect on tree health. When seedlings or small trees are severely infected, however, considerable damage or even death may result.

Figure 14. Brown felt blight caused by *Herpotrichia juniperi*. A. Lower branch infections on alpine fir. **B.** An infected branch of alpine fir. **C.** An infected branch of Engelmann spruce.

A
B
C
D

NEEDLE CASTS AND OTHER NEEDLE DISEASES OF PINE

Symptoms and Signs

Various patterns of discoloration, death, and casting of needles are the most common symptoms of this group of diseases. The main distinguishing features of each species are listed in Table 1. Positive identification of needle casts and other needle diseases can be accomplished only when mature fruiting bodies are present and can be examined microscopically, although a few species exhibit specific symptoms and can be identified easily.

Cause

Many needle casts and other needle diseases attack pine (*Pinus* spp.) and cause either premature casting or death of needles. Although casting is not always the typical symptom, 'needle cast' is the term traditionally used for a group of needle diseases caused by fungi belonging to the family Hypodermataceae (Ascomycotina: Rhytismatales). Ten common needle casts and other needle diseases of pine are listed in Table 1. All except ***Leptomelanconium cinereum*** (Dearn.) Morgan-Jones, ***Naemacyclus niveus*** (Pers.: Fr.) Sacc., and ***Thyriopsis halepensis*** (Cooke) Theissen & Sydow are needle cast fungi. ***Phaeoseptoria contortae*** Parmelee & Y. Hirat. (Deuteromycotina: Coelomycetes) and ***Hendersonia pinicola*** Wehm. (Coelomycetes) are often found on pine needles, but they are probably secondary fungi, parasitizing the needle cast fungi ***Davisomycella ampla*** (J.J. Davis) Darker (A, B) and ***Lophodermella concolor*** (Dearn.) Darker (C, D), respectively. A needle rust and the Elytroderma needle cast are discussed separately (*see* Needle Rust of Hard Pines, and Elytroderma Needle Cast of Pine).

Distribution and Hosts

Distribution and abundance of these diseases vary from year to year. Among the fungi listed in Table 1, ***Lophodermium pinastri*** (Shrad.) Chev. and ***Naemacyclus niveus*** are known on both hard and soft pines, ***Bifusella linearis*** (Peck) Hoehnel and ***Lophodermium nitens*** Darker are known only on soft pines, and all other species are known only on hard pines.

Disease Cycle

Ascospores produced on second- or third-year needles infect current-year needles in the spring or early summer. Symptoms may not show until late summer or fall. Mature fruiting bodies are usually produced during the winter and sporulate in the spring of the next year.

Damage

None of the diseases described in this chapter kill large trees or affect their health significantly unless heavy and repeated infections occur in successive years. Extensive defoliation can affect growth and shape of trees. ***Lophodermium pinastri*** causes significant damage in nurseries elsewhere, but it is not an important disease in nurseries in the prairie provinces. ***Naemacyclus minor*** Butin, which is closely related to *N. niveus*, has been reported to cause significant damage on planted Scots pine (*Pinus sylvestris* L.) in the eastern United States and has been recognized in a few Scots pine tree nurseries in Manitoba in recent years.

Figure 15. Needle casts of pine. A, B. *Davisomycella ampla* on lodgepole pine (*Pinus contorta* Dougl. var. *latifolia* Engelm.). **C, D.** *Lophodermella concolor* on lodgepole pine.

Table 1. Symptoms and signs of needle casts and other needle diseases of pine in the prairie provinces

Organism	Symptoms and signs
Bifusella linearis (Peck) Hoehnel **Needle cast**	Hysterothecia elongated, shiny black, subcuticular, of variable length on discolored (sordid) needles; ascospores elongated dumbbell-shaped (bifusiform), 41–60 × 5–7 μm
Davisomycella ampla (J.J. Davis) Darker ≡ *Hypodermella ampla* (J.J. Davis) Dearn. **Needle cast**	Hysterothecia ellipsoid, dark brown on sordid section of needles, bordered with orange-brown bands; ascospores clavate, tapering toward the base, 60–130 × 5–8 μm, gelatinous sheath 3–7 μm thick (Fig. 15A, B)
Elytroderma deformans (Weir) Darker **Elytroderma needle cast** (*see* Elytroderma Needle Cast of Pine)	Hysterothecia elongate, black, on brown second-year needles, sporulate on straw-colored third-year needles; ascospores cylindrical, 90–118 × 6–8 μm, two celled, gelatinous sheath 6–10 μm thick (Fig. 16A–D); systemic infection
Leptomelanconium cinereum (Dearn.) Morgan-Jones ≡ *Gloeocoryneum cinereum* (Dearn.) Weindl. **Needle fungus**	Black, irregular to circular cushions of spores (acervuli) under hypodermis, become erumpent, on discolored needles; conidia brown, ovoid or ellipsoid, thick-walled, verrucose, three- to six-celled, 20–25 × 6–12 μm
Lophodermella concolor (Dearn.) Darker ≡ *Hypodermella concolor* (Dearn.) Darker **Needle cast**	Hysterothecia inconspicuous, on straw-colored needles under depressions; ascospores clavate, tapering toward the base, 40–60 × 6–8 μm, gelatinous sheath 2–3 μm thick (Fig. 15C, D)
Lophodermella montivaga Petrak ≡ *Hypodermella montivaga* (Petrak) Dearn. **Needle cast**	Hysterothecia elongate, dark brown, with prominent central slits; ascospores clavate, hyaline, 40–50 × 3–4 μm, gelatinous sheath 3–4 μm thick
Lophodermium nitens Darker **Needle cast**	Hysterothecia subcuticular, shiny black, elliptical, on discolored needles; ascospores filiform, 80–120 × 2–3 μm, gelatinous sheath 2 μm thick
Lophodermium pinastri (Shrad.) Chev. **Needle cast**	Hysterothecia subepidermal, dull black, elliptical, on discolored needles; ascospores filiform, 85–140 × 1.5–2.0 μm, gelatinous sheath 2 μm thick
Naemacyclus niveus (Pers.:Fr.) Sacc. **Naemacyclus needle cast**	Ascocarps concolorous with discolored needles, elliptical, subhypodermal, open by a split; ascospores filiform, with two septa, 75–120 × 2.5–3.5 μm; conidia in hyaline pycnidia, 12–16 × 1 μm
Thyriopsis halepensis (Cooke) Theissen & Sydow **Needle spot**	Ascocarps with chlorotic spots on living needles, black, amphigenous, subcuticular, roundish; ascospores ellipsoid, two celled, hyaline to brown, 11–16 × 4–8 μm

A
B
C
D

Symptoms and Signs

In Elytroderma needle cast disease, conspicuous reddish brown discoloration begins on the first-year needles in the fall and continues to spring or early summer of their second year (A–C). Third-year needles are straw colored and have characteristic slender black fruiting bodies (hysterothecia) (D). The black fruiting bodies can also appear on reddish brown second-year needles in the fall. Positive identification of this disease is difficult when hysterothecia are not present, and it can be confused especially with symptoms caused by *Lophodermella concolor* (Dearn.) Darker. In Elytroderma needle cast, needles of the previous few years are often missing (B). Compared with the distinctive witches' broom symptom on ponderosa pine (*Pinus ponderosa* Laws.), symptoms on lodgepole and jack pines (*P. contorta* Dougl. var. *latifolia* Engelm. and *P. banksiana* Lamb.) are less conspicuous but cause distinctive thinning and distortion of the crown.

Cause

Elytroderma needle cast of pine (*Pinus* spp.) is caused by a needle cast fungus, ***Elytroderma deformans*** (Weir) Darker, belonging to the family Hypodermataceae (Ascomycotina: Rhytismatales).

Distribution and Hosts

This disease is found throughout western North America and in several locations in eastern North America (Ontario and Georgia). The main hosts are ponderosa pine and lodgepole pine, but it has also been reported on jack pine, shortleaf pine (*P. echinata* P. Mill.), pinyon pine (*P. edulis* Engelm.), and Jeffrey pine (*P. jeffreyi* Grev. & Balf.).

Disease Cycle

Once the fungus infects a tree it lives systemically in the tissue until the tree dies. Spread of the disease to different branches in the same tree often seems to occur through the stem. In the spring, ascospores produced in the fruiting bodies mature and may infect new shoots, but large numbers of new infections seem to occur only once in several years.

Damage

Damage caused by this disease on lodgepole or jack pines in the prairie provinces is probably greatly underestimated. Because of the systemic nature of the infection, tree vigor is reduced gradually over many years. When the disease affects a large number of trees in a stand, the reduction in increment is likely to be significant.

Figure 16. Elytroderma needle cast of pine, caused by *Elytroderma deformans*. A, B. Heavily infected lodgepole pine. **C.** Typical reddish brown discoloration of second-year needles and straw-colored third-year needles with black fruiting bodies. **D.** Straw-colored needles with black elongated fruiting bodies.

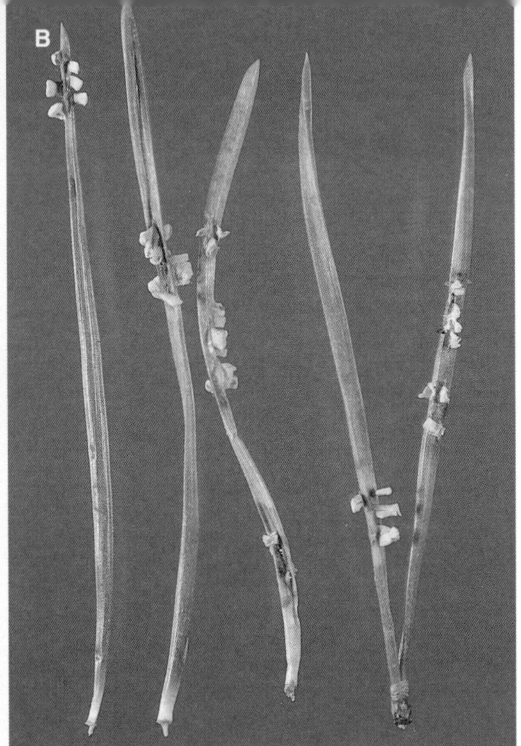

38

Symptoms and Signs

This disease can be recognized easily by the orange-yellow aeciospores produced in whitish cylindrical columns (peridermia) on second- or third-year needles in early summer (A, B). Infected needles are often paler in color and drop prematurely. On the alternate hosts, pustules of powdery, yellow urediniospores are produced on the lower side of leaves, and smooth, raised, orange-red telia appear later in the growing season (C).

Cause

A rust fungus, ***Coleosporium asterum*** (Diet.) Sydow (= *C. solidaginis* Thuemen) (Basidiomycotina: Uredinales), is the pathogen of the needle rust of hard pines (*Pinus* spp.). There are at least two different forms within the species.

Distribution and Hosts

The western form of ***C. asterum*** is known in western North America, including all the prairie provinces and the Northwest Territories, but other forms occur in eastern North America and Asia. The main hosts of this rust in the prairie provinces are lodgepole pine (*Pinus contorta* Dougl. var. *latifolia* Engelm.) and jack pine (*P. banksiana* Lamb.), but the rust also occurs on introduced Scots pine (*P. sylvestris* L.). Alternate hosts of the rust are several species of aster (*Aster* spp.) and goldenrod (*Solidago* spp.). An eastern North American species, ***C. viburni*** Arthur, occurs in Manitoba on an alternate host, viburnum (*Viburnum lentago* L.), but it has not yet been found on pine.

Disease Cycle

Infections on pine are initiated on first-year needles in late summer by basidiospores produced on alternate hosts (asters and goldenrods). The rust overwinters in infected needles and produces spermogonia and aecia the next spring. Aeciospores are disseminated by wind and infect leaves of alternate hosts. Urediniospores are then produced that can reinfect alternate hosts. Later in the season, teliospores develop on the alternate hosts and germinate to produce basidiospores.

Damage

Generally, the disease does not cause significant damage, but repeated heavy infections year after year could significantly reduce the growth of small trees. In situations such as intensively managed tree farms or nurseries, this rust has reduced the quality of trees and has been a limiting factor in the sale of ornamental or Christmas trees.

Figure 17. Needle rust of hard pines caused by *Coleosporium asterum*. A. A heavily infected branch of lodgepole pine. **B.** Close-up of infected needles with cup-shaped aecia. **C.** Uredinia and telia on an alternate host, Lindley's aster (*Aster ciliolatus* Lindl.).

A

Symptoms and Signs

In needle cast diseases, second-year needles are usually brown or straw-colored, and black fruiting bodies (hysterothecia) are produced on the discolored needles. Hysterothecia are either elongated (***Lirula macrospora*** (Hartig) Darker) (Fig. 18), fusiform (***Isthmiella crepidiformis*** (Darker) Darker), or short elliptical (***Lophodermium piceae*** (Fuckel) Hoehnel). For positive identification, ascospores produced in the fruiting bodies have to be examined. Spore measurements for the three needle casts are as follows: ***I. crepidiformis***, two celled, 60–75 × 8.0–8.5 µm; ***Lirula macrospora***, one celled, 56–68 × 2.5–3.5 µm; and ***Lophodermium piceae***, one celled, 60–95 × 1.5–2.0 µm. Fruiting bodies of ***Lophophacidium hyperboreum*** Lagerb. usually appear as circular, immersed brown patches raised into elliptical, straw-colored blisters on the underside of brownish, discolored needles. Ascospores are one celled, 15–23 × 6–8 µm.

Cause

Three needle casts of spruce (*Picea* spp.) are common in the prairie provinces: ***Isthmiella crepidiformis*** (≡ *Bifusella crepidiformis* Darker), ***Lirula macrospora*** (≡ *Lophodermium macrosporum* (Hartig) Rehm), and ***Lophodermium piceae***. 'Needle casts' is the term used for a group of needle diseases caused by fungi belonging to the family Hypodermataceae (Ascomycotina: Rhytismatales). Another fungus, ***Lophophacidium hyperboreum*** (Ascomycotina: Helotiales), which causes snow blight, is also common in the region. Spruce needle rusts and yellow witches' broom are discussed separately (*see* Needle Rusts of Spruce, and Yellow Witches' Broom of Spruce).

Distribution and Hosts

Spruce needle casts and snow blight attack all native spruces and are distributed widely in the prairie provinces.

Disease Cycle

Windborne ascospores produced on second-year needles infect newly formed needles in the spring. Symptoms may show up in the first year, but spores usually develop only on second-year needles.

Damage

Infected needles drop prematurely, usually in their second year. Severe infections in successive years affect tree growth, but needle casts do not kill the infected trees. Needle casts can be economically important in ornamental and Christmas trees because heavy infection significantly affects the appearance of trees and thus reduces their market value.

Figure 18. A needle cast of spruce caused by *Lirula macrospora*. Straw-colored, infected second-year needles of white spruce (*Picea glauca* (Moench) Voss).

A
B
C
D
E
F

Symptoms and Signs

One common symptom of infection by ***Chrysomyxa ledicola*** Lagerh., ***C. nagodhii*** P.E. Crane (formerly included in ***C. ledi*** de Bary), ***Pucciniastrum sparsum*** (Wint.) E. Fischer, and ***P. americanum*** (Farlow) Arthur (and ***P. arcticum*** Tranz.) is a slight discoloration of needles. On the needles may be found small, dotlike sexual fruiting structures (spermogonia), followed by cup- or tongue-shaped structures (aecia), which produce powdery, orange-yellow aeciospores (A–C). Infected needles drop prematurely. Position of the spermogonia on the needles (subcuticular or subepidermal) and size of aeciospores are important characteristics for species identification. ***Chrysomyxa woroninii*** Tranz. attacks only young buds, causing them to produce small, abnormal, cone-shaped or swollen branch tips (D). The spermogonia and aecia are produced on these buds. The autoecious rust ***C. weirii*** Jacks. produces tongue-shaped gelatinous telia consisting of teliospores on discolored parts of needles.

Cause

Nine needle rusts (Basidiomycotina: Uredinales) are known on various species of spruce (*Picea* spp.) in the prairie provinces (Table 2). Among them, ***Chrysomyxa ledicola*** and ***C. nagodhii*** are probably the most prevalent and most important species in the region, followed by the ***Pucciniastrum*** species. ***Chrysomyxa nagodhii*** and two less common species (***C. cassandrae*** Tranzschel and ***C. neoglandulosi*** P.E. Crane) with similar aeciospore size were formerly included under the Eurasian species ***C. ledi***; they are now considered to be separate species, based on differences in morphology and alternate hosts (P.E. Crane, Can. J. Bot. 79:957–982. 2001). Yellow witches' broom caused by *C. arctostaphyli* Diet. is discussed separately (*see* Yellow Witches' Broom of Spruce).

Distribution and Hosts

All species except ***C. weirii*** are heteroecious, that is, they need alternate hosts besides spruce to complete their life cycles. Alternate hosts for the rust species are listed in Table 2. ***Chrysomyxa weirii, C. neoglandulosi,*** and ***Pucciniastrum sparsum*** are limited mostly to the Rocky Mountain region, whereas ***C. woroninii*** occurs in mountainous and northern parts of the prairie provinces. The other needle rusts are found across the prairie provinces where native white, black, and Engelmann spruce (*Picea glauca* (Moench) Voss, *P. mariana* (Mill.) B.S.P., and *P. engelmannii* Parry) and suitable alternate hosts occur.

Disease Cycle

In host-alternating ***Chrysomyxa*** species, powdery, orange aeciospores produced on spruce needles infect young leaves of the alternate host, e.g. *Ledum* spp. (E), in the summer. Small infected areas develop on the leaves of *Ledum* in the fall. After the fungus overwinters, waxy moundlike telia form. Upon germination of teliospores, fragile basidiospores are formed and become windborne. They may infect young needles of spruce in spring. Later, orange uredinia are formed on the same leaves of *Ledum* spp. (F), and powdery, orange urediniospores produced in them can reinfect leaves of *Ledum* spp. The life cycles of ***Pucciniastrum*** spp. are similar to those of *C. ledicola* and *C. nagodhii*, except that telia overwinter on dead plants on the ground rather than on live leaves. When ***C. woroninii*** infects *Ledum* spp., small witches' brooms are produced and the fungus can survive on them perennially.

Damage

Infection by needle rusts, especially ***C. ledicola*** and ***C. nagodhii***, occasionally can be very heavy, and all or most of the current-year needles can be affected and dropped prematurely. Heavy infections seldom occur in successive years, however, and trees do not seem to be damaged significantly. Heavy infection of ornamental and Christmas trees significantly reduces their value.

Figure 19. Needle rusts of spruce. A, B. Heavy infection of *Chrysomyxa ledicola* on white spruce. **C.** Close-up of *C. ledicola* infected needles of white spruce showing ridges of aecia. **D.** Close-up of deformed buds caused by *C. woroninii*. **E.** Common Labrador tea (*Ledum groenlandicum* Oeder), one of the main alternate hosts of *C. ledicola*, *C. nagodhii*, and *C. woroninii*. **F.** Uredinia and telia of *C. ledicola* on the upper surface of common Labrador tea.

Table 2. Symptoms and signs of needle rusts of spruce in the prairie provinces

Organism	Telial hosts[a]	Symptoms and signs on spruce
Chrysomyxa arctostaphyli Diet. **Spruce broom rust** (*see* Yellow Witches' Broom of Spruce)	Kinnikinnick (*Arctostaphylos uva-ursi* (L.) Spreng.); III only, on underside of leaf (Fig. 20D)	Perennial, produces conspicuous witches' broom; spermogonia subepidermal; aeciospores 16–25 × 23–35 µm (Fig. 20A–C)
Chrysomyxa cassandrae Tranzschel **Spruce–leatherleaf rust**	Leatherleaf (*Chamaedaphne calyculata* (L.) Moench); II, III on underside of leaf	Causes premature defoliation; spermogonia subepidermal; aeciospores 14–22 × 17–31 µm with elongated smooth spots
Chrysomyxa ledicola Lagerh. **Large-spored spruce–Labrador tea rust**	Labrador tea; II, III on upper side of leaf (Fig. 19F)	Causes premature defoliation; spermogonia subepidermal; aeciospores 14–32 × 23–42 µm (Fig. 19C)
Chrysomyxa nagodhii P.E. Crane **Small-spored spruce–Labrador tea rust**	Labrador tea (*Ledum* spp.); II, III on underside of leaf	Causes premature defoliation; spermogonia subepidermal; aeciospores 14–24 × 15–31 µm with elongated smooth spots
Chrysomyxa neoglandulosi P.E. Crane **Spruce–glandular Labrador tea rust**	Glandular Labrador tea (*Ledum glandulosum* Nutt.); II, III on underside of leaf	Causes premature defoliation; spermogonia subepidermal; aeciospores 14–22 × 18–25 µm
Chrysomyxa weirii Jacks. **Weir's spruce cushion rust** (Autoecious species)	Spruce (*Picea* spp.); III on needles	Orange-yellow tongue-like telia on discolored bands on second-year needles
Chrysomyxa woroninii Tranz. **Spruce shoot rust**	Labrador tea (causes witches' broom); III on systemically infected leaf	Attacks young buds and causes small cone-shaped swelling; spermogonia subepidermal; aeciospores 17–35 × 24–62 µm (Fig. 19D)
Pucciniastrum americanum (Farlow) Arthur and *P. arcticum* Tranz. **Spruce–raspberry rust**	Raspberry (*Rubus* spp.); II, III on underside of leaf	Causes premature defoliation; spermogonia subcuticular; aeciospores 13–16 × 16–26 µm
Pucciniastrum sparsum (Wint.) E. Fischer **Spruce–bearberry rust**	Alpine bearberry (*Arctostaphylos rubra* (Rehder & Wils.) Fern.); II, III on underside of leaf	Causes premature defoliation; spermogonia subcuticular; aeciospores 13–22 × 20–32 µm with elongated smooth spots

[a] II = uredinial state; III = telial state.

A
B
C
D

Symptoms and Signs

This is the most conspicuous disease of spruce (*Picea* spp.) in the prairie provinces. Yellowish witches' brooms of different sizes can be recognized from a distance (A, B). Needles of the affected brooms die and are shed in the fall, leaving bare branches in winter and early spring. In the spring, systemically infected needles emerge on the brooms; small, dotlike spermogonia are produced on them, followed by orange-yellow, blisterlike aecia (C). Affected needles are pale yellow. Witches' brooms increase in size every year, and big brooms 20 to 30 years old are common. Witches' brooms are also produced by other causes, for example, physiological or dwarf mistletoe, but they are green and not conspicuously yellow like this disease. On the alternate host, kinnikinnick or common bearberry (*Arctostaphylos uva-ursi* (L.) Spreng.), deep-orange-colored telial sori are produced on the purple-brown spots on the underside of the leaves (D).

Cause

Yellow witches' broom of spruce is caused by a rust fungus, ***Chrysomyxa arctostaphyli*** Diet. (Basidiomycotina: Uredinales).

Distribution and Hosts

Yellow witches' broom is known on all native and introduced spruces across North America, but it is more common in the west. The alternate host of this rust is kinnikinnick or common bearberry.

Disease Cycle

Powdery, yellow airborne aeciospores produced on spruce needles in early summer are carried to the current-year leaves of kinnikinnick and initiate infection. During the next spring, teliospores are produced on the underside of the second-year leaves. Urediniospores are not produced. Fragile basidiospores are produced in the spring as a result of teliospore germination and they infect young shoots of spruce. Infections become systemic, and after several years of abnormal proliferation of short shoots, witches' brooms are produced. After spermogonia appear, aecia are produced every year on systemically infected needles produced on the broom.

Damage

Yellow witches' broom of spruce affects the shape of the crown, produces dead tops, and sometimes results in tree mortality. Damage caused by the disease in the prairie provinces has not been appraised but is considered insignificant. In some parts of the United States, this disease is considered to increase incidence of decay by providing entry points for pathogens.

Figure 20. Yellow witches' broom of spruce, caused by *Chrysomyxa arctostaphyli*. A. Distant view of a yellow witches' broom on white spruce (*Picea glauca* (Moench) Voss). **B.** A young broom on white spruce. **C.** Close-up of infected spruce needles showing spermogonia (blackish dots) and aecia (orange-yellow ridges). **D.** Alternate state (telia) on the lower surface of kinnikinnick.

A

B

Symptoms and Signs

Symptoms and signs of needle diseases of fir vary depending on the species of pathogen and stage of the disease. Generally, discoloration and premature death of needles and the kind of fruiting structures present are the important criteria for identification. Microscopical examinations of certain spore states is necessary for definite identification; collections made during midsummer often lack mature spores, which makes identification difficult. Distinguishing characteristics are listed in Table 3.

Cause

Eight common needle casts and other needle diseases of fir (*Abies* spp.) that occur in the prairie provinces are listed in Table 3. They all belong to the subdivision Ascomycotina. Needle rusts of fir are discussed separately (*see* Needle Rusts of Fir).

Distribution and Hosts

Two fir species occur in the prairie provinces: subalpine fir (*Abies lasiocarpa* (Hook.) Nutt.), which occurs from western Alberta to British Columbia, particularly at higher elevations in the Rocky Mountains; and balsam fir (*A. balsamea* (L.) Mill.), found in the boreal forest zone of all three prairie provinces. The distribution of needle diseases of these two fir species within the region is not well known.

Disease Cycle

In general, windborne ascospores produced on overwintered diseased needles, either on the tree or on the ground, infect newly formed needles in early summer. Fruiting structures may be visible during the growing season (A, B), but spores usually do not mature until the next spring.

Damage

Fir needle casts and other needle diseases seldom kill trees; however, severe infection can reduce the vigor of small trees and consequently make them vulnerable to other pathogens. Heavy infection by any needle disease is undesirable for ornamental trees.

Figure 21. A needle cast on alpine fir caused by *Isthmiella abietis*. A. Straw-colored, infected second-year needles. **B.** Close-up of infected needles showing dark, continuous fruiting bodies (hysterothecia) along midribs.

Table 3. Symptoms and signs of common needle casts and other needle diseases of fir in the prairie provinces

Organism	Symptoms and signs
Isthmiella abietis (Dearn.) Darker ≡ *Bifusella abietis* Dearn. **Needle cast**	Hysterothecia on discolored needles, continuous along midrib, black; ascospores bifusoid, one celled, hyaline, 40–50 × 4–6 µm, with gelatinous sheath, eight spores per ascus (A, B)
Isthmiella quadrispora Ziller **Needle cast**	Hysterothecia on yellow, dead needles, continuous along midrib, black; ascospores bifusiform to biclavate, one celled, 40–55 × 3–6 µm, with gelatinous sheath, four spores per ascus
Lirula abietis-concoloris (Mayr ex Dearn.) Darker ≡ *Hypodermella abietis-concoloris* Mayr ex Dearn. **Needle cast**	Hysterothecia on discolored needles, black or bluish black lustre, continuous along midrib; ascospores clavate, hyaline, one celled, 70–104 × 4–5 µm, with gelatinous sheath
Lophomerum autumnale (Darker) Magasi **Needle cast**	Hysterothecia scattered on both sides of needles, shiny black, roundish; ascospores filiform, up to four celled, 85–95 × 1.5–2.0 µm, with gelatinous sheath
Nothophacidium abietinellum (Dearn.) J. Reid & Cain **Snow blight**	Disk-shaped apothecia, up to 0.8 mm in diameter, develop subepidermally and become exposed, on discolored needles; ascospores oval to broad-ellipsoid, 5.5–8.5 × 4.5–6.5 µm, hyaline, occasionally brown with age
Phacidium abietis (Dearn.) J. Reid & Cain **Snow blight**	Disk-shaped apothecia, up to 0.5 mm in diameter, arranged in linear series on each side of the midrib, intrahypodermal on discolored needles; ascospores ellipsoid to fusiform, curved, hyaline, 17–25 × 5–7 µm
Phaeocryptopus nudus (Peck) Petrak **Needle fungus**	Globose, black fruiting bodies (pseudothecia), up to 1 mm, scattered on the surface of discolored needles; ascospores slightly clavate, hyaline to pale brown, two celled, 10–15 × 4–5 µm
Sarcotrochila balsameae (J.J. Davis) Korf ≡ *Stegopezizella balsameae* (J.J. Davis) Sydow **Snow blight**	Disk-shaped apothecia, up to 0.5 mm in diameter, on the underside of discolored needles; ascospores ovoid, hyaline, first one celled, 10–25 µm, becoming brown and four celled, 24–30 × 7–10 µm

A
B
C
D
E
F

Symptoms and Signs

Infected needles of fir (*Abies* spp.) are often chlorotic or discolored (yellow, brown) or become chlorotic and shed prematurely. Small, dotlike spermogonia and yellowish orange or white cup-shaped aecia (C) with powdery aeciospores are produced on the undersides of needles. ***Melampsorella caryophyllacearum*** Schroet. causes conspicuous witches' broom symptoms (E) similar to yellow witches' broom of spruce.

Cause

Six species of fir needle rust (Basidiomycotina: Uredinales) are common in the prairie provinces: ***Hyalopsora aspidiotus*** (Magn.) Magn., ***Melampsora abieti-capraearum*** Tub., ***Pucciniastrum epilobii*** Otth (A, B), ***P. goeppertianum*** (Kuehn) Kleb. (C, D), ***Uredinopsis phegopteridis*** Arthur, and ***Melampsorella caryophyllacearum*** (E, F).

Distribution and Hosts

Two species of fir occur in the prairie provinces: alpine fir (*Abies lasiocarpa* (Hook.) Nutt.) and balsam fir (*A. balsamea* (L.) Mill.). Distribution of each rust species within the region is not well known. All species of fir needle rust are heteroecious, that is, they need alternate hosts to complete their life cycles. Alternate hosts for each species are as follows:

H. aspidiotus—oak fern (*Gymnocarpium dryopteris* (L.) Newm.);
Melampsora abieti-capraearum—willow (*Salix* spp.);
P. epilobii—fireweed (*Epilobium angustifolium* L.) (B);
P. goeppertianum—blueberry and huckleberry (*Vaccinium* spp.) (D) and cranberry (*V. vitis-idaea* L. var. *minus* Lodd.);
U. phegopteridis—oak fern (*G. dryopteris*);
Melampsorella caryophyllacearum—chickweed (*Cerastium* spp. and *Stellaria* spp.).

Disease Cycle

The typical disease cycle of a fir needle rust, exemplified by ***P. epilobii***, is as follows. The pathogen overwinters in cushion-shaped fruiting bodies (telia) on dead leaves of fireweed. In the spring, teliospores germinate to produce fragile, windborne basidiospores. They infect newly produced fir needles. In a few weeks minute, dotlike sexual fruiting structures (spermogonia) are produced. They are followed by cup-shaped aecia containing powdery, yellow aeciospores. These spores infect leaves of fireweed and produce a repeating spore state (uredinia) with powdery urediniospores; later, cushion-shaped, dark-colored telia are formed. The disease cycles of other needle rusts of fir are similar to that of *P. epilobii.* In ***M. caryophyllacearum***, however, infection on fir is perennial. It survives systemically on expanding witches' broom tissues. Spermogonia and aecia are produced annually on newly formed needles on the witches' broom.

Damage

Only ***P. epilobii*** is known to cause any significant damage in the prairie provinces. Young alpine fir (*Abies lasiocarpa* (Hook.) Nutt.) are often heavily infected with this species. All current-year needles can be infected, and growth of trees is apparently inhibited if heavy infections occur in successive years. Heavily infected trees are usually not killed, however, and recover from the effect of the rust in a few years. After trees reach 2 m or more, impact of the rust becomes insignificant. Two species, ***P. goeppertianum*** and ***M. caryophyllacearum***, are known to cause economic damage to Christmas tree plantations in the maritime provinces. The western form of ***P. goeppertianum*** that exists in the Rocky Mountains is different from the eastern form and does not cause any damage to needles because of the slow development of aecia.

Figure 22. Needle rusts of fir. A, B. *Pucciniastrum epilobii.* **A.** Heavily infected alpine fir. **B.** Infected alpine fir and fireweed, the alternate host, showing uredinia and telia. **C, D.** *P. goeppertianum.* **C.** Infection on alpine fir, showing elongated hornlike aecia. **D.** Swollen, brown, infected stems with telia and an uninfected alternate host (far right), grouseberry (*Vaccinium scoparium* Leiberg). **E, F.** *Melampsorella caryophyllacearum.* **E.** A witches' broom caused by the rust on balsam fir. **F.** Close-up of infected branchlets with ridges of yellow aecia.

INSECTS AND DISEASES OF CONIFERS

Roots, Stems, and Branches

A
B
C
D
E
F

Symptoms and Signs

The mountain pine beetle, ***Dendroctonus ponderosae*** Hopkins (Coleoptera: Scolytidae), and associated blue stain fungi act together to kill trees. The blue stain fungi (Ascomycotina: Ophiostomatales) most consistently present and vectored by *D. ponderosae* are ***Ophiostoma clavigerum*** (Robins.-J. & Davids.) Upad., ***O. minus*** (Hedgc.) Syd. & P. Syd., and ***O. huntii*** (Robins.-J.) De Hoog & R.J. Scheffer. These fungi are believed to stop water transport in the stem and kill infected trees. Accumulations of pitch or sawdust (C) are conspicuous around entrance holes bored into the bark of trees by adult beetles from mid-July to mid-August. Sawdust is quickly blown or washed away but pitch tubes may remain for more than a year after attack. Needles of trees attacked in midsummer do not fade until the next spring, and turn reddish brown by summer (F). During the fall and winter after attack, woodpeckers feed on bark- and wood-boring insects on infested trees. Trunks of trees foraged on by woodpeckers are reddish brown instead of the normal grayish brown and have piles of bark fragments at their base. Removal of bark from infested trees reveals galleries, adults, and larvae. Egg galleries are 10–41 cm (average, 28 cm) long and have a short basal diagonal section followed by a much longer vertical section (D). Adults are black, stout-bodied, cylindrical, and 4.0–7.5 mm long (A). Larvae are creamy-white, legless grubs with light brown heads (E), and are 6–7 mm long when fully grown. Grayish blue staining of sapwood (B) is the most conspicuous symptom of colonization of ray parenchyma cells by blue stain fungi. Various fungal fruiting structures (such as synnemata and perithecia) and mycelium of blue stain fungi and other fungi are often evident in beetle galleries and pupal chambers.

Distribution and Hosts

The mountain pine beetle is found throughout the southern half of British Columbia and western Alberta, in the Cypress Hills of southeastern Alberta and southwestern Saskatchewan, and south to New Mexico and California. The main host is lodgepole pine (*Pinus contorta* Dougl. var. *latifolia* Engelm.) but any other species of pine planted within the range of the mountain pine beetle can be attacked.

Life Cycle

Adults emerge and attack green trees in midsummer. They construct vertical egg galleries in the phloem (D) and lay eggs along the sides. Larvae feed away from the egg galleries until early fall, overwinter, and continue feeding in the spring. Pupation occurs in late spring to early summer, and new beetles feed for a few days before emerging to attack new trees. Adults transport blue stain fungi to new trees by carrying spores on their surface and within a specialized sac (mycangium) in the mouth area.

Damage

The mountain pine beetle is the most serious enemy of mature pines in western Canada. Outbreaks of this insect occurred in Banff National Park, Alberta, from 1940–44, and in southwestern Alberta from 1977–85. It has been estimated that over 1 million m^3 of lodgepole pine was killed in Alberta during the most recent outbreak. The monetary loss may be far greater than indicated by the volume loss because high-value mature trees are preferred by the beetles. In addition to direct volume loss, outbreaks of mountain pine beetle upset harvesting plans, reduce aesthetic value in recreational areas, and increase fire hazard. Blue stain and extensive checking of sapwood from salvaged trees killed by mountain pine beetles lowers the commercial value of trees used for lumber and pulp.

Figure 23. Mountain pine beetle, *Dendroctonus ponderosae*. A. Adult. **B.** Sapwood of lodgepole pine logs infested with blue stain fungi. **C.** Pitch tubes caused by beetles boring into the stem. **D.** Adult egg galleries (vertical) and larval galleries (horizontal). **E.** Larva. **F.** Beetle-killed lodgepole pine.

A
B
C
D

E

Symptoms and Signs

The first sign of attack by adult spruce beetles, ***Dendroctonus rufipennis*** (Kirby) (Coleoptera: Scolytidae), is the presence of entrance holes bored into the bark of the lower stem of trees in late May and June. Conspicuous boring dust accumulates on the bark scales below these holes (*see* Fig. 25F, Other Bark Beetles) and on the ground at the base of the tree. The boring dust becomes less conspicuous toward fall because of the action of wind and rain. Pitch (pitch tubes) may also accumulate around the entrance holes of standing healthy trees (*see* Fig. 23C, Mountain Pine Beetle). During attack the foliage of infested trees looks normal. Larval feeding and colonization by associated fungi eventually disrupt the phloem transport system, thereby girdling the tree. The foliage begins to fade to a yellowish green during the winter after attack (E). The trees tend to lose most of their needles by the next fall. During the fall and winter after attack, woodpeckers remove bark from infested trees as they search for beetle larvae and adults to eat. Trunks of trees foraged on by woodpeckers are reddish instead of the normal gray. This is most noticeable in early spring. Removal of bark from infested trees reveals galleries, adults, and larvae (C, D). Egg galleries are 6–23 cm long (average, 13 cm) and follow the grain of the wood. Adults are stout-bodied, cylindrical insects that are 4.5–7.0 mm long. Newly emerged adult beetles have a dark pronotum and black or reddish brown elytra (A). Larvae are creamy-white, legless grubs with light brown heads (C), and they are 6–7 mm long when fully grown.

Distribution and Hosts

This species has a transcontinental distribution and attacks principally white spruce (*Picea glauca* (Moench) Voss) and Engelmann spruce (*P. engelmannii* Parry) in the prairie provinces.

Life Cycle

The life cycle of this species may require 1–3 years, depending on latitude, altitude, and exposure. In the prairie provinces most beetles have a 2-year life cycle. Only beetles that have overwintered as adults are capable of reproduction the following summer. Adults attack new hosts in late spring and early summer. The female bores through the bark and constructs short egg galleries in the phloem, engraving the sapwood (D). The galleries tend to follow the grain of the wood. Eggs are deposited in clumps along the sides of the egg galleries (B). Larvae tend to feed in groups in the early instars and individually as they mature. Overwintering larvae resume development in the spring, pass through a brief pupal stage, and transform into adults. Adults emerge from standing trees in August and September, move to the base of the trees, reenter the bark, and hibernate. Adults in windfall or slash remain under the bark and do not emerge until the next spring.

Damage

The spruce beetle is capable of attacking and killing standing, healthy timber (E), especially if large numbers of beetles are present after breeding in wood killed or weakened by fires, windstorms, logging operations, or defoliation by other insects. Such outbreaks of spruce beetle in standing timber have been reported on several occasions in northern and southwestern Alberta.

Figure 24. Spruce beetle, *Dendroctonus rufipennis*. A. Adult. **B.** Egg gallery and eggs. **C.** Larvae. **D.** Adult and larval galleries. **E.** White spruce killed by spruce beetles.

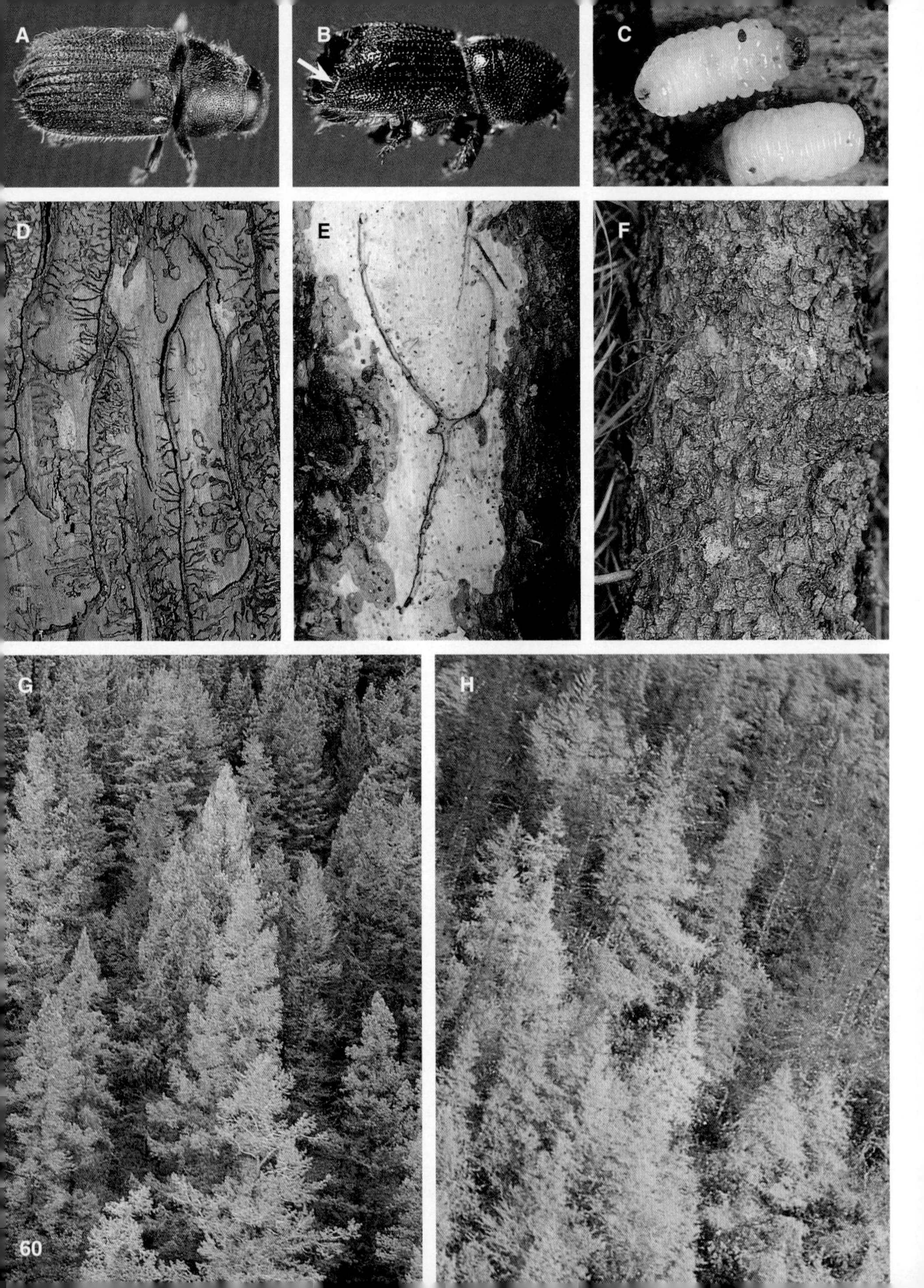
A
B
C
D
E
F
G
H

Symptoms and Signs

The eastern larch beetle, ***Dendroctonus simplex*** LeConte (Coleoptera: Scolytidae); Douglas-fir beetle, ***D. pseudotsugae*** Hopkins; and pine engraver, ***Ips pini*** (Say) (Scolytidae), attack trees in late spring by boring through the bark into the phloem. This activity produces sawdust (rarely pitch tubes), which accumulates on bark scales below the entrance holes (F) and at the base of trees. In tamarack, ***D. simplex*** entrance holes are often marked by copious resin flow. The foliage of larch killed by ***D. simplex*** usually turns yellow (H) in August, a few weeks ahead of foliage of healthy trees. The foliage of trees killed by the two ***Dendroctonus*** begins to fade (G) in late summer. Egg galleries of ***I. pini*** (E) consist of a central nuptial chamber and two to five radial tunnels. Egg galleries of ***D. simplex*** (D) are sinuous, vertical tunnels that are often branched; galleries of ***D. pseudotsugae*** are straighter and less branched than those of *D. simplex*. Larvae of all three species are creamy-white, legless grubs with light brown heads (C), and are 4–5 mm long when fully grown. Adults are stout, cylindrical insects. ***Ips*** adults have a concave elytral declivity, which has spines on the lateral margins (B). ***Ips pini*** adults are dark brown and 3.5–4.5 mm long. Adult ***D. simplex*** (A) are 3.5–5.0 mm long, have a spineless, convex elytral declivity, and are typically reddish brown (some are blackish); ***D. pseudotsugae*** adults look similar but are slightly larger. Other species of *Ips* and *Dendroctonus* also occur in the prairie provinces. Identification of bark beetles is difficult and usually requires expert assistance.

Distribution and Hosts

Dendroctonus simplex has a transcontinental boreal distribution. It attacks tamarack (*Larix laricina* (Du Roi) K. Koch), and exotic larches. ***Dendroctonus pseudotsugae*** occurs from western Alberta and British Columbia south to northern Mexico; it attacks Douglas-fir (*Pseudotsuga menziesii* (Mirb.) Franco). ***Ips pini*** occurs throughout North America and attacks mainly pines (*Pinus* spp.), but sometimes spruces (*Picea* spp.).

Life Cycle

Dendroctonus simplex adults overwinter under the bark, emerge in May and June, attack new trees, and construct galleries (D) and lay eggs in the phloem of the bole. Larvae tunnel laterally away from the main galleries (D). Pupation occurs in the bark. New adults appear in August; most stay under the bark to feed and overwinter, but some emerge, move to the base of the trees, and reenter the bark to overwinter. One generation is produced per year, but some adults may produce up to three separate broods in different trees. The life cycle of ***D. pseudotsugae*** is similar to that of *D. simplex*, and one or two broods may be produced per year.

Ips pini adults overwinter in the duff. Males attack trees in the spring and construct nuptial chambers in the phloem. After females arrive and mate, each constructs a separate egg gallery, which radiates outward from the nuptial chamber (E). Some females may emerge after laying eggs, reenter the bark elsewhere, and lay a second batch of eggs. Development of immature stages is similar to that of *D. simplex*. ***Ips pini*** has one or two generations per year.

Damage

All three beetle species are capable of killing healthy trees in large numbers, but damage in the prairie provinces has not been extensive. Small outbreaks of ***D. simplex*** have occurred in Saskatchewan and Manitoba. Small infestations of ***D. pseudotsugae*** have occurred in the foothills and mountains of Alberta. ***Ips pini*** attacks mainly dying and recently dead trees, but sometimes kills standing trees around the margins of clear-cuts.

Figure 25. Other bark beetles. A. Adult *Dendroctonus simplex*. **B.** Adult *Ips* species (arrow points to elytral declivity). **C.** *D. simplex* larvae. **D.** Adult (vertical) and larval (horizontal) galleries of *D. simplex*. **E.** Adult galleries of *Ips pini*. **F.** Sawdust on pine trunk resulting from attack by *I. pini*. **G.** Douglas-fir killed by *D. pseudotsugae*. **H.** Tamarack killed by *D. simplex*.

A
B
C
D
E
F
G

Symptoms and Signs

Larvae of the Warren root collar weevil, ***Hylobius warreni*** Wood (Coleoptera: Curculionidae), and pine root collar weevil, ***H. radicis*** Buchanan, feed on phloem in the root collar area and lateral roots of conifers ranging from 5 years old to maturity. Resin flowing from damaged bark mixes with soil and hardens to form a whitish, heavy encrusted layer over the feeding sites (D). Removal of this encrusted layer reveals old girdling (D, F) and new girdling caused by feeding larvae. Larvae (C) are creamy-white, legless grubs with reddish brown heads. Mature ***H. warreni*** and ***H. radicis*** larvae are 18–20 mm long and 15–18 mm long, respectively. Adults cause little damage but may sometimes be observed feeding around the root collar area. Adults of ***H. warreni*** (A) are stout, wingless, 12–15 mm long, with heads prolonged into snouts. They are mostly black but are covered in whitish or pale yellow scales that form irregular spots on the elytra and give the body a grayish appearance. Adults of ***H. radicis*** (B) are 10–12 mm long and dull reddish brown, and have yellow scales that form spots on the elytra. The Couper collar weevil, ***H. pinicola*** (Couper), is similar in appearance to *H. warreni*, but has wings.

Distribution and Hosts

Hylobius warreni has a transcontinental boreal distribution and feeds primarily on white spruce (*Picea glauca* (Moench) Voss), jack pine (*Pinus banksiana* Lamb.), and lodgepole pine (*P. contorta* Dougl. var. *latifolia* Engelm.). ***Hylobius radicis*** occurs from Nova Scotia to southeastern Manitoba. In Manitoba, it attacks Austrian pine (*P. nigra* Arnold), Scots pine (*P. sylvestris* L.), red pine (*P. resinosa* Ait.), and jack pine. ***Hylobius pinicola*** has a transcontinental distribution and feeds on spruces (*Picea* spp.), tamarack (*Larix laricina* (Du Roi) K. Koch), and possibly firs (*Abies* spp.).

Life Cycle

Hylobius warreni adults feed mainly on the phloem of branches and twigs. Eggs are laid around the root collar from May to September (peak in early July). Larvae confine their feeding to portions of the root collar and lateral roots that are between the mineral soil horizon and the surface. Larval development requires about 2 years and is completed in late spring. Mature larvae move a few centimeters away from the feeding site and construct hardened pupal cells (E) from soil, bark chips, and resin. After pupation, most adults emerge in August and September (some emerge the next spring). Adults may require overwintering before they can reproduce. Because adults live for up to 4 years and reproduce each year, there may be three or more overlapping generations, and all stages can be found from late spring to early fall. The life cycle and habits of ***H. radicis*** and ***H. pinicola*** are similar to those described above.

Damage

Larval feeding may completely girdle and kill trees (G); young trees are more likely to die than older trees. Partial root collar girdling may reduce lateral root growth, which adversely affects the stability of the root system and may lead to blowdown. Damaged trees are also susceptible to attack by opportunistic insects and fungi. Trees that survive several years of feeding damage form callus tissues, which may overgrow the wounds, and the root collar becomes fluted. ***Hylobius warreni*** is most abundant on moist, well-drained, highly productive growing sites. It causes significant damage in natural and managed stands. In southeastern Manitoba, ***H. radicis*** is mainly a pest of trees in plantations and is particularly abundant on sites with well-drained sandy soils.

Figure 26. Root collar weevils. A. Adult *Hylobius warreni*. **B.** Adult *H. radicis*. **C–G.** *H. warreni*. **C.** Larva feeding in phloem. **D.** Lodgepole pine partially girdled by feeding larvae. Note adjacent resin accumulations in the soil. **E.** Pupa and adult in pupal cells. **F.** Larval feeding wounds on an old lodgepole pine. **G.** Young lodgepole pine killed by larval feeding.

A
B
C
D
E
F

Symptoms and Signs

The presence of adults of the whitespotted sawyer beetle, ***Monochamus scutellatus*** (Say) (Coleoptera: Cerambycidae), flying around dying and recently dead trees and logs is often the first sign of attack. Adults (A) have lateral spines on the thorax, are black or mottled gray except for a white spot at the base of the elytra, have long antennae, and are 20–25 mm long. Larvae feed in the phloem and score the surface of the wood (D), mostly from July onwards. Other wood borers do not score the wood surface to the same extent as *Monochamus*. When larvae bore into the wood in August, they chew characteristic oval-shaped entrance holes (D), which are 6–8 mm in diameter along the longest axis, larger than those of other common wood borers. The wood chips resulting from larval boring are expelled from the galleries and accumulate in piles at the base of infested trees (B) or logs (C). When adults leave the wood, they chew circular holes 8–11 mm in diameter, much larger than those caused by other common wood borers. Larvae (E) are creamy-white grubs with a yellowish thorax and brown head. Larvae are up to 20–25 mm long while in the phloem, but are 35–50 mm long when fully grown. ***Monochamus*** larvae differ from those of most other common wood-boring insects by their larger size and lack of legs; also the sides of the head are roughly parallel, whereas other species commonly have heads with rounded or diverging sides. The northeastern sawyer, ***M. notatus*** (Drury), causes similar signs and symptoms. Adults (F) are mottled gray, white, and black and are 25–35 mm long. Larvae are similar to, but 10–15% larger than, those of *M. scutellatus*.

Distribution and Hosts

Both species have transcontinental distributions. ***Monochamus scutellatus*** attacks most native and introduced pines (*Pinus* spp.), spruces (*Picea* spp.), and true firs (*Abies* spp.), as well as Douglas-fir (*Pseudotsuga menziesii* (Mirb.) Franco) and tamarack (*Larix laricina* (Du Roi) K. Koch); ***M. notatus*** attacks eastern white pine (*Pinus monticola* Dougl.), balsam fir (*Abies balsamea* (L.) Mill.), and spruces.

Life Cycle

Monochamus scutellatus usually has a 2-year life cycle in the prairie provinces; however, adults are produced each year. Adults fly from late May to August. They feed on the bark and twigs of live coniferous trees a few days before seeking out suitable breeding material. Eggs are laid in June and July into slits chewed in the bark by females. Eggs hatch in about 2 weeks. Larvae feed in the cambium and phloem for several weeks before entering the sapwood, where they feed until late fall. Feeding resumes the next spring and continues throughout the summer, when extensive damage is done in the sapwood and heartwood. Larvae overwinter a second time and pupate the next spring in cells constructed near the wood surface. Adults emerge in late spring. ***Monochamus notatus*** has a similar life cycle.

Damage

Sawyer beetles seriously degrade logs stored for too long in the forest or mill yard. They also shorten the time that logs can be salvaged for lumber after fires or death caused by insect outbreaks. The impact is especially serious if lumber with 'worm holes' is destined for European and Asian markets, where it is feared that the pine wood nematode (***Bursaphelenchus xylophilus*** (Steiner & Buhrer)), which is vectored by ***Monochamus***, may be accidentally introduced. Sawyer beetles have little impact on pulp wood. When abundant, adult ***M. scutellatus*** may cause serious feeding damage to live trees.

Figure 27. Sawyer beetles. A–E. *Monochamus scutellatus.* **A.** Adult. **B.** Wood chips and frass at base of an infested dead spruce. **C.** Wood chips and frass caused by larval feeding in log decks. **D.** Excavations in surface of sapwood made by larva. Note oval entrance hole at the left. **E.** Larvae and galleries in white spruce log. **F.** *M. notatus* adult.

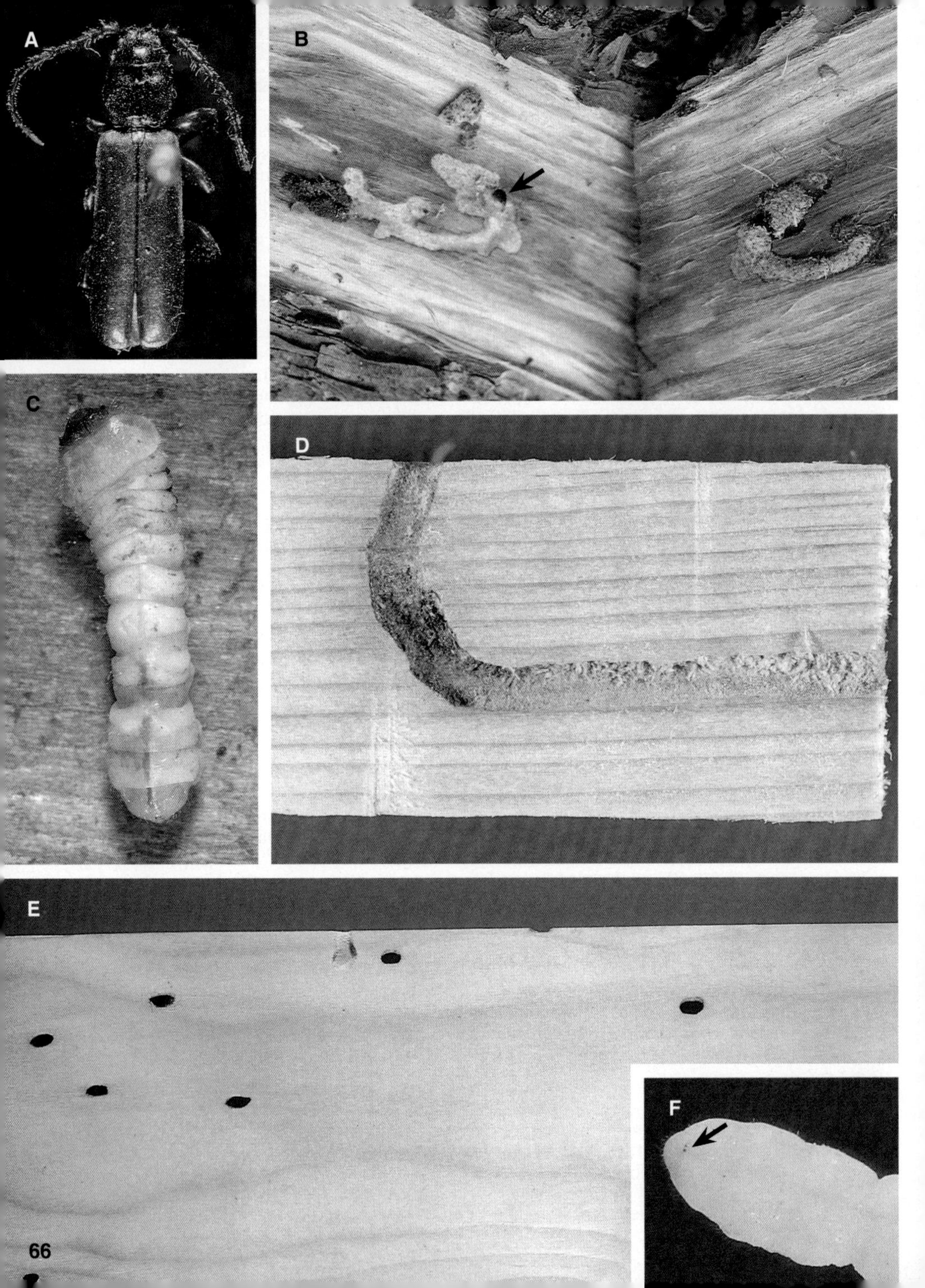
A
B
C
D
E
F

Symptoms and Signs

The northern spruce borer, ***Tetropium parvulum*** Casey (Coleoptera: Cerambycidae), attacks dying and recently dead trees and logs. As larvae feed in the phloem they lightly score the surface of the wood (B, left). Boring dust and frass are packed in the galleries behind the feeding larvae (B, right); very little is expelled to the surface. When larvae move into the wood in late July and August, they construct oval entrance holes (B) that are 4–6 mm in diameter, along the longest axis. Larvae are grublike and creamy-white, except for a yellowish thorax and brown head (C). They have three pairs of rudimentary thoracic legs and are 15–20 mm long when fully grown. Larvae can be distinguished from those of other wood borers by the presence of two light brown peglike structures, visible with a 10× hand lens, on the dorsal part of the last abdominal segment (F). When larvae enter the sapwood, they excavate characteristic L-shaped galleries that penetrate the wood 25–35 mm before they turn at a right angle to become parallel with the wood grain (D). A related species, ***T. cinnamopterum*** Kirby, is very similar to *T. parvulum*, and causes similar signs and symptoms.

Distribution and Hosts

Tetropium parvulum and ***T. cinnamopterum*** both have a boreal transcontinental distribution and attack white (*Picea glauca* (Moench) Voss) and Engelmann spruce (*P. engelmannii* Parry); however, ***T. cinnamopterum*** also occasionally attacks balsam fir (*Abies balsamea* (L.) Mill.).

Life Cycle

Both species have similar life cycles. Adults emerge from infested logs in May and June. Adults are 8–12 mm long, brown to brownish black, and the pronotum is usually slightly darker than the elytra (A). Females mate and lay eggs under bark scales. Eggs hatch in about 12 days. On hatching, larvae feed in the bark and cambium (B) for about 6 weeks before entering the sapwood in late July and August, where they excavate L-shaped galleries (D). Larvae excavate pupation chambers at the distal end of the galleries, plug the entrance to the chambers with wood chips, and overwinter. Pupation occurs the following spring. The pupal stage lasts about 10 days, and newly emerged adults remain in the pupal chamber for about another 10 days before emerging through the larval entrance hole and chewing through the bark.

Damage

These species show a preference for large diameter (>45 cm) logs but will also attack smaller logs. Larval mines in the sapwood (E) degrade lumber. Larvae penetrate the wood to a depth of 25–35 mm, but occasionally up to 75 mm. Because of this shallow boring habit, most of the larval mines are usually eliminated with the slabs and edgings. Larvae are also associated with wood-staining fungi, which discolor the wood surrounding the larval tunnels and contribute to lumber degrade.

Figure 28. Northern spruce borer, *Tetropium parvulum*. A. Adult. **B.** White spruce tree with bark removed to expose a larval feeding gallery in the bark and cambium and an entrance hole (see arrow) into the wood. Note the boring dust packed into the gallery. **C.** Larva. **D.** Section of lumber showing the shape of a larval gallery in the sapwood. **E.** Section of lumber showing holes made by larvae. **F.** Posterior dorsal part of the abdomen of a larva showing the two peglike structures on the last segment (arrow).

A
B
C
D
E
F

Symptoms and Signs

Several species of horntails in the genera ***Sirex***, ***Urocerus***, and ***Xeris*** (Hymenoptera: Siricidae) infest only dying or recently dead coniferous trees and logs. Adults are wasplike insects, 9–30 mm long, and females have a long ovipositor at the rear end of the body (A). The basic body color is bluish black but many species have orange markings. Eggs are laid directly into the sapwood, hence tunneling does not occur in the phloem. Larvae make long, meandering tunnels (C) primarily in the sapwood. Tunnels are up to 40 cm long and filled with fine, tightly packed frass (C). Larvae have yellowish heads, creamy-white bodies, small thoracic legs, and a chitinous dorsal horn at the rear of the abdomen (B).

Adults of the red and black carpenter ant, ***Camponotus herculeanus*** (L.) (Hymenoptera: Formicidae), excavate decayed portions and exposed heartwood of live trees, stumps and logs, and sometimes man-made wooden structures, to make nests. They do not eat wood. Excavation is evident by the piles of coarse sawdust (E), holes in the wood, and the movement of ants through the holes. These ants do not make individual galleries but excavate large, hollow chambers (F). Adult ants (D) are dark colored and 6–20 mm long. Workers are not winged, but males and females have two pairs of membranous wings. Females shed their wings after dispersal. Other species of carpenter ants in the genus ***Camponotus*** are also present in the prairie provinces.

Distribution and Hosts

The family Siricidae has a transcontinental distribution; individual species may have less wide distributions. Most species of native and exotic spruces (*Picea* spp.), pines (*Pinus* spp.), larches (*Larix* spp.), and true firs (*Abies* spp.) as well as Douglas-fir (*Pseudotsuga menziesii* (Mirb.) Franco) are attacked by horntails. ***Camponotus herculeanus*** has a transcontinental distribution and will infest any species of conifer that is suitable for nest construction.

Life Cycle

The life cycles of most horntail species are poorly known, but are likely similar to that of the blue horntail, ***Sirex cyaneus*** (Fabricius). This species requires 2–3 years to complete development. Adults emerge during the summer and early fall. Females use their ovipositor to puncture the sapwood and lay eggs. Some eggs hatch immediately but others overwinter. Larvae feed mainly in the sapwood and seldom penetrate deeper than 4 cm. Galleries are up to 20 cm long. Larvae pass through 5–11 instars, and overwinter once or twice. Pupal cells (C) are formed near the surface of the wood, and adults chew through the bark to emerge.

Winged adults of ***C. herculeanus*** mate and disperse in the spring. Mated females (queens) shed their wings and seek suitable locations in which to start new colonies. Each queen initially rears 10–20 workers by feeding them from food reserves in her body. Subsequent enlargement and maintenance of the colony is done by workers. Most eggs develop into workers; the rest develop into winged males and females, which disperse and reproduce the next spring. Colonies may be active as long as 15 years.

Damage

Horntail galleries reduce the quality of lumber, but volume loss is minimal because the tunnels are near the surface. Some horntails are known to disseminate wood decay fungi. Carpenter ant nests constructed in decayed living trees may further weaken trees and contribute to breakage during windstorms. Carpenter ants also attack wood in service; they may infest parts of buildings and of power and telephone poles where excessive moisture accumulates. This may weaken wooden structures, and ants from nests in and around houses are a public nuisance.

Figure 29. Horntails and carpenter ants. A–C. *Sirex cyaneus*. **A.** Adult. **B.** Larva. **C.** Larval gallery and pupal chamber (arrow). **D–F.** *Camponotus herculeanus*. **D.** Wingless adult. **E.** Living spruce tree containing ant nest. Note pile of wood chips at base of tree. **F.** Cross section of spruce showing nest.

A
B
C
D
E
F
G
H
I

Symptoms and Signs

In the spring, the first signs of attack by white pine weevil, ***Pissodes strobi*** (Peck) (Coleoptera: Curculionidae), are the presence of adults and resin beads (C) on last year's leader. Adults (A) are a mottled brown color with variable white and yellow patches on their backs, their heads have long snouts, and they are 5–8 mm long. Resin oozes from punctures (C) caused by adults inserting their mouthparts through the bark of the leader and feeding on the phloem. Later, feeding larvae in the phloem girdle the leader and cause the current growth to wilt (G) in early to mid-June and then turn yellowish brown (H) by July. Removal of bark from fading terminals reveals larvae (D), which are legless grubs with creamy-white, wrinkled bodies and reddish brown heads; they are about 10 mm long when fully grown. Mature larvae excavate cavities in the wood and line them with wood chips (E). Pupation occurs in these 'chip cocoons'. Emerging adults make small circular holes, 2–3 mm in diameter, in the bark (F). Old previously infested terminals (I) may remain attached to the tree for 15 years, and may be identified by the presence of feeding punctures, chip cocoons, or adult emergence holes.

Distribution and Hosts

The white pine weevil has a transcontinental distribution. In the prairie provinces this species attacks primarily white spruce (*Picea glauca* (Moench) Voss) and Engelmann spruce (*P. engelmannii* Parry), but sometimes also attacks other species of native and exotic pines (*Pinus* spp.) and spruces.

Life Cycle

Adults overwinter in the duff and emerge in early spring (March–April), often before all snow has disappeared from the ground. They move up the trees to the previous year's leaders, feed on the phloem, and deposit eggs (B) in the feeding punctures from late April to mid-May. After hatching, larvae feed gregariously in the phloem, moving down the terminal (D). Larvae feed for 5–6 weeks and pass through four instars. Mature larvae excavate chip cocoons (E) in the outer wood, in which they pupate. New adults emerge from infested terminals from late July to early September, feed for several weeks on healthy branches, and then overwinter in the duff. There is one generation per year.

Damage

This species attacks trees in natural stands but prefers open-growing (low density, no shade), vigorous ornamental trees, plantations, and nursery stock that are 1.5–8.0 m high. Larval feeding always kills at least 2 years growth (current and previous year's) (H) and sometimes 3 or more, depending on how far down the main shoot the larvae tunnel. After the leader dies, one or more of the side branches nearest to the top turns upward and assumes leadership, resulting in a crooked or forked tree (I) with a bushy top. Secondary organisms such as heartwood rot fungi may enter through weevil-killed leaders. Small trees <0.5 m tall may be killed by weevils in a single year. This insect is one of the most destructive pests of young spruce plantations in western Canada. In Alberta, up to 20% of white spruce may be infested each year in some plantations. Severely affected plantations may be commercially valueless.

Figure 30. White pine weevil, *Pissodes strobi*. A. Adult. **B.** Egg. **C.** Adult feeding punctures and resulting resin beading on white spruce terminal. **D.** Larvae feeding in phloem of white spruce. **E.** Mature larva and chip cocoons (image rotated 90°). **F.** Adult emergence holes in white spruce terminal (image rotated 90°). **G.** Wilting terminal shoots of white spruce caused by weevil attack. **H.** White spruce leader killed by weevils. **I.** Old weevil damage on white spruce and resulting crooking and branching of the main stem (note remnants of old infested terminal at arrow).

A
B
C
D
E
F
G
H

Symptoms and Signs

The first signs of attack by the lodgepole terminal weevil, ***Pissodes terminalis*** Hopping (Coleoptera: Curculionidae), are the presence of adults, feeding punctures (B), and resin bleeding on the current year's terminal leaders. Adults (A) are a mottled brown color with variable white and yellow patches on their elytra, their heads have long snouts, and they are 5–9 mm long. These signs start to appear in May in jack pine (*Pinus banksiana* Lamb.) and June in lodgepole pine (*P. contorta* Dougl. var. *latifolia* Engelm.), and are visible only with careful examination of leaders. Later, when larvae feed in the phloem, the bark over the feeding tunnels turns magenta (D) and often bleeds resin. Removal of bark from the discolored portions of phloem reveals young larvae (C). Later, larvae enter the pith of the terminal and are easily visible if the terminal is split open (E). Larvae are legless grubs with creamy-white, wrinkled bodies and reddish brown heads, and are 10–12 mm long at maturity. Infested jack pine terminals start to fade to yellowish brown and curl at the top in July (H). Adults emerge from jack pine in late July and August, creating circular emergence holes, 2–4 mm in diameter. Fading of infested lodgepole pine terminals (G) begins in September; they eventually turn a brick-red color over the winter but do not curl. Examination of the pith of discolored terminals during winter reveals overwintering larvae and frass (E). Adults emerge from lodgepole pine terminals the following mid-July to late August. No chip cocoons are produced by this species and pupation occurs in the pith (F).

Distribution and Hosts

This species occurs from Manitoba to British Columbia and the Yukon Territory, and ranges south to California. It attacks lodgepole pine and jack pine.

Life Cycle

In western Alberta, adults overwinter in the duff, emerge in late May and June, move up the trees to the current year's leader, feed on the phloem, and deposit eggs in feeding punctures in June and early July. Larvae feed in the phloem (C), towards the terminal bud, until about mid-August, when as third and fourth instar larvae, they enter the pith. Larvae feed until about late September, overwinter, continue feeding the next spring, and pupate (F) in late June. New adults emerge between mid-July and late August, feed for several weeks, overwinter, and emerge the next spring to reproduce. Because the whole population is not synchronized new adults are produced and new trees are attacked each year. In jack pine stands, *P. terminalis* has a 1-year life cycle. Adults emerge from the duff between May and June and reproduce. Adults of the next generation emerge from terminals from mid-August until September, feed, and move to the duff to overwinter.

Damage

This species attacks trees in natural stands but prefers open-growing trees (i.e., low density and little shade), 2–9 m in height, in plantations. Repeated attacks on the same tree cause a crooked or forked stem and bushy crown, thus reducing the tree's value as lumber. Leader loss can be recovered in 2–3 years if the tree is not attacked in succeeding years. The yearly incidence of attacks is typically 2–5%, but may be as high as 30%. Cumulative incidence of attacks may be as high as 87%.

Figure 31. Lodgepole terminal weevil, *Pissodes terminalis*. A. Adult. **B.** Adult feeding punctures (see arrows) on lodgepole pine terminal. **C.** Young larva in phloem. **D.** Discolored bark on lodgepole pine terminal caused by larval feeding in the phloem. **E.** Mature larvae overwintering in the pith of a lodgepole pine terminal. **F.** Pupae in the pith of a lodgepole pine terminal. **G.** Fading needles on lodgepole pine terminal caused by weevil attack. **H.** Fading needles on jack pine terminal caused by weevil attack.

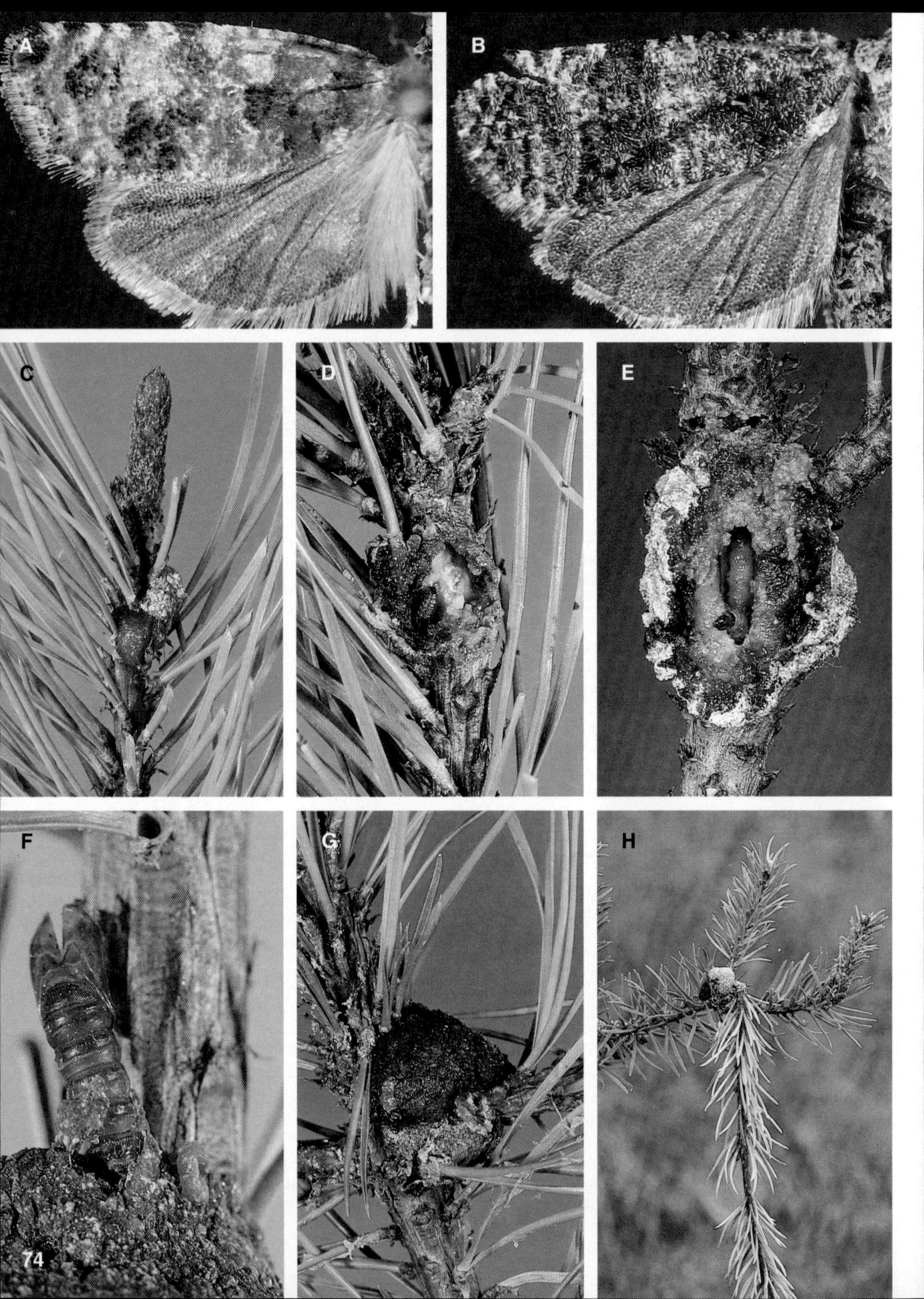
A
B
C
D
E
F
G
H
74

Symptoms and Signs

Larvae of the northern pitch twig moth, ***Petrova albicapitana*** (Busck) (Lepidoptera: Tortricidae), and the metallic pitch blister moth, ***P. metallica*** (Busck), bore into the shoots, twigs, branches and stems of pines, and construct characteristic blisters (up to 20 mm in diameter) from resin (C, G). Larvae feed in the cavities underneath blisters (D, E). Feeding larvae often girdle twigs and branches, which results in discoloration of the foliage distal to the blisters and breakage at the blisters (H). Larvae of ***P. metallica*** feed in the xylem and pith, tunnelling downward beyond the nodule and causing a slight swelling of the shoot (E); ***P. albicapitana*** larvae feed in the bark and outer xylem, and do not cause shoot swelling (D). Blisters caused by ***P. metallica*** usually have a rough surface with whitish areas of dried resin (E); those of ***P. albicapitana*** are generally smooth and unicolorous (G). Mature larvae of both species are yellowish to orange-brown, have reddish heads and thoracic shields, are 15–17 mm long, and have thoracic legs and five pairs of abdominal prolegs.

Distribution and Hosts

Petrova albicapitana occurs from Nova Scotia to British Columbia and the Northwest Territories. The primary hosts are lodgepole pine (*Pinus contorta* Dougl. var. *latifolia* Engelm.) and jack pine (*P. banksiana* Lamb.). ***Petrova metallica*** occurs from British Columbia and western Alberta to southern Yukon, and in the Cypress Hills of southeastern Alberta. Its main hosts are lodgepole pine and ponderosa pine (*P. ponderosa* Laws.). Where the ranges of the two species overlap, ***P. metallica*** occurs at higher elevations (above 800 m). Both species of *Petrova* attack young to mature trees, but most injuries occur on 0.5- to 5.0-m-tall trees growing in natural and planted stands.

Life Cycle

Petrova albicapitana has a 2-year life cycle; moths are produced every year (June to July) because populations are not synchronized. Moths have mottled reddish brown fore wings and a 14- to 21-mm wingspan (A). Eggs are laid near the bases of terminal buds. Each young larva excavates a cavity in the cortex tissue, begins feeding, and spins a layer of silk over the cavity, which is combined with pitch and frass to form a small blister (C). Larvae overwinter in the blisters, resume feeding the next spring, and enlarge the cavities and blisters. By early June, larvae vacate the first feeding sites, migrate down the branches, and establish second feeding sites, usually at the base of branch whorls (G). They continue feeding during the second summer and enlarge the new blisters and cavities. After overwintering a second time, larvae feed briefly before pupation (F) and moth emergence.

The life cycle of ***P. metallica*** is less understood, but appears to require 1–2 years, depending on climate. Adults are similar in size to *P. albicapitana*, but have dark metallic-gray, mottled fore wings (B). Each larva of this species excavates only one feeding site.

Damage

Feeding of ***P. albicapitana*** may weaken or kill the terminal leader, resulting in stem deformities and height growth reduction. Partly girdled stems are prone to break by wind or snow for up to 5 years after injury. At high population levels, 60–70% of trees in 8- to 12-year-old lodgepole pine stands may be attacked per year. In the upper third of the tree crown, up to 33% of the shoots may be killed in a single year. Damage caused by ***P. metallica*** is not as serious because it is usually limited to twigs.

Figure 32. Pitch blister moths. A. Adult *Petrova albicapitana*. **B.** Adult *P. metallica*. **C.** Early-instar *P. albicapitana* blister. **D.** *P. albicapitana* larva and feeding damage (blister surface removed). **E.** *P. metallica* larva at entrance to tunnel in pith (blister surface removed). Note the slight swelling of the stem around the blister. **F.** Empty pupal case of *P. albicapitana* protruding from blister. **G.** Late-instar *P. albicapitana* blister. **H.** Lateral of pine broken following *P. albicapitana* feeding.

A
B
C
D
E
F
G
H
I

Symptoms and Signs

Larvae of the eastern pine shoot borer, ***Eucosma gloriola*** Heinrich (Lepidoptera: Tortricidae); yellow jack pine shoot borer, ***Rhyacionia sonia*** Miller (Tortricidae); and jack pine shoot borer, ***R. granti*** Miller, bore into pine shoots to feed, causing girdling. Infested shoots have slower needle growth than healthy shoots. Before needles of infested shoots fade conspicuously, most larvae have completed development and left the shoots. ***Eucosma gloriola*** feeding tunnels are straight and follow the longitudinal axis of the shoot, and frass is packed at the ends of the tunnel rather than expelled. ***Rhyacionia*** feeding tunnels are more crooked (H), and some frass is expelled and adheres to the outside of the shoot by silken webbing. ***Eucosma gloriola*** prefers leaders; ***Rhyacionia*** infests laterals and leaders. The needles of twigs infested by ***E. gloriola*** and ***R. sonia*** grow to near maturity before they die (C, I) and fade; needles of twigs infested by ***R. granti*** remain immature (F). Larvae of ***E. gloriola*** (B) have dark brown heads, brownish thoracic shields, and pale brown bodies, and are about 13 mm long at maturity. Mature larvae of ***R. granti*** (F) have reddish brown bodies and dark brown heads, and are about 15 mm long. Larvae of ***R. sonia*** (G) are yellowish brown with dark brown heads, and are about 9 mm long at maturity. All have thoracic legs and five pairs of abdominal prolegs.

Distribution and Hosts

All three species occur from eastern Canada to southeastern Manitoba and in adjacent parts of the United States. ***Eucosma gloriola*** attacks mainly eastern white pine (*Pinus strobus* L.), but also other native and exotic pines. Only jack pine (*P. banksiana* Lamb.) is attacked in Manitoba. ***Rhyacionia sonia*** and ***R. granti*** attack mainly jack pine (I).

Life Cycle

Eucosma gloriola adults emerge from late May to early June. Adults have a wingspan of 13–16 mm, the fore wings are copper-red with two irregular, gray bands (A). Eggs are laid on the developing shoots. Young larvae bore into the pith and tunnel towards the base of the shoot. There is usually one larva per shoot (as many as six in heavy infestations). Frass is not expelled but is packed tightly behind the larvae. Larval tunnels are 10–29 mm long. Fifth-instar larvae reverse direction and tunnel upward for several centimeters before girdling the shoot. After boring for an additional 5–10 cm, larvae chew exit holes through the bark, leave the shoots, enter the litter, spin silken cocoons that incorporate soil debris, pupate therein, and overwinter.

Rhyacionia granti adults emerge from late May to early June, and ***R. sonia*** adults emerge about a month later. Adults of ***R. granti*** have a wingspan of 15–20 mm and alternating grayish and brownish bands on the fore wings (D); ***R. sonia*** adults are similar (E) but have a wingspan of about 12 mm. Eggs are laid on the shoot and larvae construct tunnels in the xylem, leaving only the bark (H). In late July and August, mature larvae cut exit holes in the bark, through which they exit to pupate and overwinter in the soil.

Damage

All three species have damaged young jack pine plantations in southeastern Manitoba. Girdling weakens shoots so that they wilt or break during high winds. Shoots that do not break eventually recover, but growth is usually distorted. ***Eucosma gloriola*** prefers leaders, and repeated attacks may cause ‘cabbage-topped’ trees. Severe damage to Christmas trees may make them unmarketable during the year of attack.

Figure 33. Pine shoot borers. A–C. *Eucosma gloriola*. **A.** Adult. **B.** Larva in shoot. **C.** Current and previous (see stub at arrow) jack pine leaders broken as a result of larval feeding. **D.** Wings of *Rhyacionia granti* adult. **E.** Wings of *R. sonia* adult. **F.** Jack pine shoot killed by *R. granti* larva. **G.** *R. sonia* larva. **H.** Jack pine shoot tunnelled by *R. granti* larva. **I.** Jack pine shoot killed by *R. sonia*.

A
B
C
D
E
F
G
H
78

Symptoms and Signs

These insects suck sap from the bark of trees and cause foliage discoloration (E) and premature needle drop in severe infestations. The first symptom of large aphid infestations is often the presence of sticky honeydew on and around infested trees. Often the honeydew becomes infected with sooty mold, which gives the branches a blackish appearance. The most common aphids on conifers are those in the genus ***Cinara*** (Homoptera: Aphididae), which feed openly in colonies on the bark. They are soft-bodied, 1–5 mm long, and usually darkly colored (A), and do not produce flocculence. Hard pine adelgids, ***Pineus coloradensis*** (Gillette) (Homoptera: Adelgidae), are covered by white flocculence on the bark and needles of current growth (C). They have soft, oval, slate-gray bodies that are 0.5–2 mm long (D). Female pine tortoise scales, ***Toumeyella parvicornis*** (Cockerell) (Homoptera: Coccidae), are oval, strongly convex, reddish brown with black markings, and about 6 mm long when fully grown (F); they usually occur on small twigs and branches. Female spruce bud scales, ***Physokermes piceae*** (Schrank) (Coccidae), are about 3 mm long, reddish brown, and hemispherical, resembling buds (H); they are clustered near the bases of small twigs.

Distribution and Hosts

The genus ***Cinara*** has a transcontinental distribution and feeds on most species of conifer; individual species have more limited geographical and host ranges. ***Pineus coloradensis*** occurs from western Alberta and British Columbia to Colorado and California, and feeds on lodgepole pine (*Pinus contorta* Dougl. var. *latifolia* Engelm.) and other pines. ***Toumeyella parvicornis*** occurs from the east coast to southeastern Manitoba. Its principal host is jack pine (*P. banksiana* Lamb.), but Scots pine (*P. sylvestris* L.) and red pine (*P. resinosa* Ait.) are also attacked. ***Physokermes piceae*** is an introduced species that occurs on spruces (*Picea* spp.) from the east coast to Alberta and the Northwest Territories.

Life Cycle

Cinara aphids overwinter as eggs attached to the foliage (B) or twigs. Eggs hatch in the spring, and up to five parthenogenetic generations occur throughout the summer. Most aphids are wingless, but some winged females are produced, which disperse to new hosts before reproducing. A sexual generation of winged males and females occurs in late summer or early fall. After mating, females deposit eggs that overwinter. Little is known about the life history of ***P. coloradensis***. Wingless parthenogenetic adults secrete white waxy filaments that cover their bodies. They apparently reproduce indefinitely on the same host. Winged adults are known, but no alternate hosts have been recorded.

Toumeyella parvicornis has one generation per year and overwinters as half-grown scales on branches. Development resumes in mid-May and new nymphs emerge from the genital aperture of mature females in late June. These tiny reddish nymphs (crawlers) (G) disperse to find feeding sites. Male scales are small, elongate, and whitish, and emerge in late August to mate with immature females. Females feed until fall and overwinter. ***Physokermes piceae*** has a life cycle similar to that of *T. parvicornis*, except that immature males overwinter attached to needles, and complete their development and mate with females in the spring.

Damage

These insects predominate on young trees in natural stands and plantations and on ornamental trees. Infestations may cause foliage discoloration (E), leading to premature needle drop. Although trees may be attacked in consecutive years, there is generally little adverse effect. Loss of needles and flocculence caused by ***P. coloradensis*** may reduce the value of Christmas trees. Severe infestations of ***T. parvicornis*** can kill branches or young trees in 1–2 years.

Figure 34. Aphids, adelgids, and scales. A, B. *Cinara*. **A.** Aphids on lodgepole pine. **B.** Eggs on pine needles. **C, D.** *Pineus coloradensis*. **C.** Flocculence on lodgepole pine shoot. **D.** Adult with flocculence removed. **E–G.** *Toumeyella parvicornis*. **E.** Infested jack pine. **F.** Females feeding on jack pine. **G.** Crawlers. **H.** *Physokermes piceae* on spruce.

A

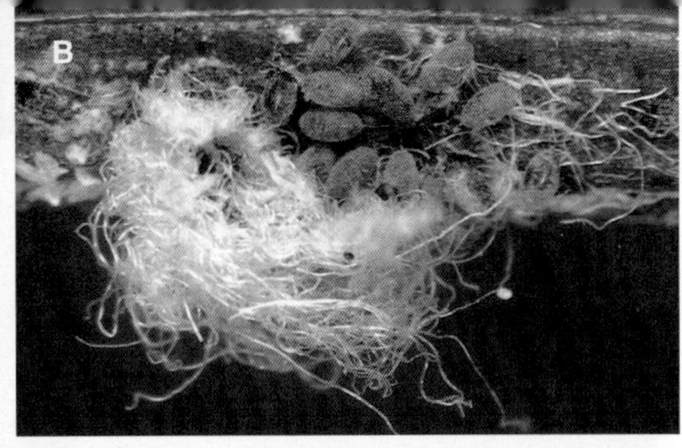
B

C

D

E

F

G

Symptoms and Signs

The Cooley spruce gall adelgid, ***Adelges cooleyi*** (Gillette) (Homoptera: Adelgidae); spruce gall adelgid, ***A. lariciatus*** (Patch); pale spruce gall adelgid, ***A. strobilobius*** (Kaltenbach); and ragged spruce gall adelgid, ***Pineus similis*** (Gillette) (Adelgidae), cause cone-shaped galls to form on the new shoots of spruces in late spring. Galls are swellings 2–6 cm long that contain hundreds of adelgids feeding on sap. Initially the galls are green but later some turn reddish purple. Old vacated galls have mouthlike openings and are reddish brown (D, G). Old galls may remain on branches for several years. In ***Adelges*** galls (D, F, G), the needles are fused together at their bases, whereas in ***Pineus*** galls (E), the bases of the needles are flattened and scalelike but are not fused together. The galls of ***A. cooleyi*** (C, D) are usually 4–6 cm long and those of the other two ***Adelges*** are 2–4 cm long and pineapple-shaped (F, G). Galls formed by ***A. strobilobius*** are at the ends of the shoots and usually surround the shoots, and those caused by ***A. lariciatus*** often do not surround the shoot or have part of the shoot extending beyond the tip of the gall; however, it is generally difficult to differentiate between these two species. The presence of adelgids is also indicated by white cottony specks (flocculence) (A) that appear on infected trees in the spring and summer. Adelgid nymphs (B) and adults are small, dark, soft-bodied insects.

Distribution and Hosts

Adelges lariciatus and ***A. strobilobius*** are distributed from Alberta to eastern Canada and in adjacent parts of the United States. Both species alternate between spruces (*Picea* spp.) and tamarack (*Larix laricina* (Du Roi) K. Koch). ***Adelges cooleyi*** is transcontinental but is most abundant in the west where its alternate host, Douglas-fir (*Pseudotsuga menziesii* (Mirb.) Franco), is present. ***Pineus similis*** is transcontinental; it has no alternate host.

Life Cycle

Adelges cooleyi has a 2-year life cycle. In the summer, winged parthenogenetic females leave Douglas-fir and migrate to spruce, where they lay eggs that develop into small, wingless males and females. These insects then mate, and each female lays a single egg on spruce. Emerging nymphs feed on spruce needles and in the fall move to stem tips, just below the buds, where they overwinter. The next spring nymphs develop into parthenogenetic females, which each lay about 200 eggs. Nymphs emerging from these eggs move to new growth to feed, causing the formation of galls. The winged parthenogenetic females that later emerge from the galls migrate to Douglas-fir to lay their eggs. The nymphs hatching from these eggs overwinter on the foliage. These nymphs develop into winged and wingless forms in the following spring. The winged females migrate to spruce to start the cycle again. There is also a wingless parthenogenetic form of ***A. cooleyi*** that causes flocculence on spruce and is capable of reproducing without need of an alternate host. ***Adelges lariciatus*** and ***A. strobilobius*** have a life cycle similar to that of *A. cooleyi*.

Pineus similis is a parthenogenetic species, has two generations per year, and overwinters as young nymphs near the tips of twigs. Feeding of these nymphs in the spring stimulates gall formation. Adult females lay eggs that develop into winged and wingless females. These females migrate and produce eggs that develop into nymphs that overwinter.

Damage

Adelgids mainly affect tree appearance; old galls remain on branches for several years, making trees unsightly. If gall formation is heavy, growth and vigor may be reduced, but trees are rarely killed.

Figure 35. Adelgid galls on spruce. A–D. *Adelges cooleyi*. **A.** Flocculence on foliage caused by adelgids. **B.** Nymphs under flocculence. **C.** Young gall. **D.** Mature gall. **E.** Galls caused by *Pineus similis*. **F, G.** *A. lariciatus*. **F.** Young gall. **G.** Mature gall.

A

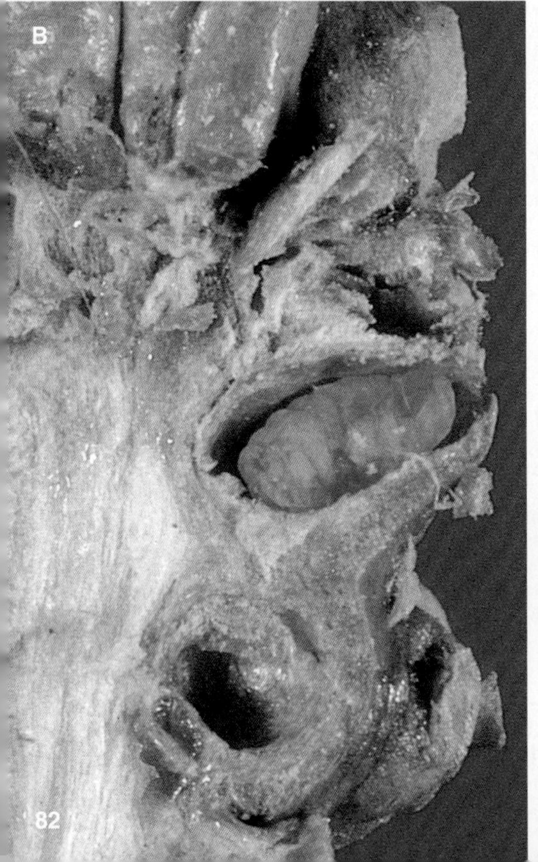
B
82

C

Symptoms and Signs

The spruce gall midge, ***Mayetiola piceae*** (Feltham) (Diptera: Cecidomyiidae), causes gall formation on twigs. Galled portions of twigs are up to twice the diameter of uninfested portions and 3–6 cm long. Most remain the same color as the bark of unaffected twigs, but some have a reddish tinge (A). Spruce gall midge larvae are seen if infested galls are cut open. Mature larvae are legless, 1.3–1.7 mm long, and reddish orange (B). Old abandoned galls have conspicuous emergence holes (C) and may remain on the tree for several years.

Distribution and Hosts

This insect species is distributed from eastern Canada to Alberta and the Yukon and in adjacent parts of the United States. It attacks white spruce (*Picea glauca* (Moench) Voss).

Life Cycle

The spruce gall midge overwinters as mature larvae in the galls. Pupation occurs in the spring and adults emerge from the galls formed in the previous year's shoot in late spring, leaving conspicuous emergence holes (C). Adults are small, reddish brown, two-winged flies. Females lay eggs at the base of needles on the new shoots. After hatching, larvae bore into the shoots and form cells. Galls are formed by the swelling of tissue around the larval cells.

Damage

Spruce gall midge infestations cause the affected shoots to curl and often to die. Large infestations of this species occurred in the Yukon in the 1960s and in northern Alberta and the Northwest Territories in the 1990s. During outbreaks as much as 100% of current shoots may be infested on some trees in a given year.

Figure 36. Spruce gall midge, *Mayetiola piceae*. A. Galling of spruce twig caused by midge larvae. **B.** Cross section of a spruce shoot showing a larva in its cell. **C.** Adult emergence holes in galls.

A
B

C

D

E

Symptoms and Signs

The most conspicuous symptom of dwarf mistletoe infection is the production of witches' brooms (A, B). The size and shape of the brooms vary depending on the host species and the age and position of the infection. Witches' brooms on jack pine (*Pinus banksiana* Lamb.) (A) tend to be much larger and more open than those on lodgepole pine (*P. contorta* Dougl. var. *latifolia* Engelm.) (B). Infected branches are usually twisted and produce spindle-shaped swellings (C). Dwarf mistletoe is dioecious (having separate male and female plants). Aerial shoots of dwarf mistletoe are greenish yellow and up to 10 cm long and consist of branch segments with reduced scalelike leaves. Depending on the sex of the plant, they produce male (D) or female flowers during late spring or early summer. After pollination, berries are produced on female plants in the summer and remain on the plant for about a year before they ripen (E). When aerial shoots become nonfunctional, they fall off and leave basal cups, which are connected to the endophytic system of the dwarf mistletoe in the pine tissue. Basal cups are important diagnostic features of dwarf mistletoe because they remain for several years; witches' brooms and stem swellings can result from many other causes, and fresh aerial shoots are often absent.

Cause

Lodgepole pine dwarf mistletoe (***Arceuthobium americanum*** Nutt.) is a parasitic flowering plant belonging to the family Viscaceae. A related dwarf mistletoe (eastern dwarf mistletoe, ***A. pusillum*** Peck) commonly occurring on black spruce (*Picea mariana* (Mill.) B.S.P.) and white spruce (*P. glauca* (Moench) Voss) in Manitoba and the eastern edge of Saskatchewan is discussed separately (*see* Eastern Dwarf Mistletoe).

Distribution and Hosts

Distribution of lodgepole pine dwarf mistletoe in Canada extends from the Coast Range in British Columbia to the southeast shore of Lake Winnipeg in Manitoba. The two main hosts of the pathogen are lodgepole pine and jack pine, but it is also occasionally found on white spruce (*Picea glauca* (Moench) Voss) and Scots pine (*Pinus sylvestris* L.) as well as several other pine species (*P. albicaulis* Engelm., *P. edulis* Engelm., *P. flexilis* James, *P. jeffreyi* Grev. & Balf., and *P. ponderosa* Laws.).

Disease Cycle

The single, large seed in a berry is forcefully discharged during mid-August to mid-September. The seed is surrounded by a hygroscopic, sticky material called viscin, which enables it to adhere to the surface on which it lands. When seeds land on susceptible host tissue they overwinter, germinate in May or June, and penetrate the host. Young twigs are most susceptible. It takes about 3–5 years before the infection produces aerial shoots. Flower and seed production usually begins 1–2 years after the shoots first appear.

Damage

This disease causes the greatest amount of annual loss in merchantable lodgepole and jack pines in the prairie provinces. Heavily infected young trees will not reach merchantable size. Dwarf mistletoe not only causes mortality but also reduces growth by about one-third. Dwarf mistletoe is particularly damaging to immature stands up to 50 years old. The unsightly appearance of infected trees due to brooming is also a serious concern in townsites and parks.

Figure 37. Lodgepole pine dwarf mistletoe, *Arceuthobium americanum*. A. Heavily infected jack pine. **B.** Two heavily infected lodgepole pines with conspicuous witches' broom symptom. **C.** Aerial shoots of dwarf mistletoe and spindle-shaped swelling of infected stem of lodgepole pine. **D.** Close-up of male flowers. **E.** Mature berries.

A

B

Symptoms and Signs

The most obvious symptom of mistletoe infection on black spruce is the presence of witches' brooms, bushy, compact masses of branches and twigs (A). Presence of aerial shoots or basal cups of the parasite are the most positive signs of the mistletoe infection. Aerial shoots are green to brown, unbranched, and up to 3 cm high, and they have scalelike leaves (B). They distinguish this disease from witches' brooms caused by a rust fungus, ***Chrysomyxa arctostaphyli*** Diet. (*see* Yellow Witches' Broom of Spruce). Also, brooms caused by the mistletoe are green, whereas those of yellow witches' broom are yellow. This mistletoe, like lodgepole pine dwarf mistletoe, is dioecious (having separate male and female plants).

Cause

Eastern dwarf mistletoe, ***Arceuthobium pusillum*** Peck, is a parasitic flowering plant belonging to the family Viscaceae.

Distribution and Hosts

Distribution of eastern dwarf mistletoe in Canada extends from Newfoundland westward to eastern Saskatchewan. It also occurs in Minnesota and northern Pennsylvania in the United States. The main hosts of the pathogen are black spruce (*Picea mariana* (Mill.) B.S.P.) and white spruce (*P. glauca* (Moench) Voss). Other known but occasional or rare hosts include Colorado blue spruce (*P. pungens* Engelm.), eastern larch (*Larix laricina* (Du Roi) K. Koch.), jack pine (*Pinus banksiana* Lamb.), red pine (*P. resinosa* Ait.), and eastern white pine (*P. strobus* L.).

Disease Cycle

The life cycle of eastern dwarf mistletoe from infection to the first seed production requires at least 4 years. A single seed in a fruit matures in the fall and is forcefully discharged; if it lands on susceptible host tissue (host tissue less than 5 years old), it overwinters and penetrates into the host in the spring. It takes at least 4 years before shoot and seed formation. Flowers of both sexes appear in the spring and seeds mature in the fall.

Damage

Both black spruce and white spruce are significantly damaged. The damage includes reduced growth, poor wood quality, and poor seed production. The mistletoe kills young trees and over several years may kill older trees (A). In heavily infested stands, stocking levels are so low that a commercial harvest is impossible.

Figure 38. Eastern dwarf mistletoe, *Arceuthobium pusillum*. A. Witches' brooms on dying white spruce. **B.** Brown mistletoe shoots on white spruce twigs.

A

B

C
CFRN
entrance

D

Symptoms and Signs

Trees with burls often occur in clusters, and multiple burls are common on a tree. Globose or semiglobose galls are produced on the main stems and branches of spruce (A, B) and pine. Gall tissues are usually sound, having no decay or stain; branch or stem death distal to burls does not occur. For this reason, many spruce burls have been used ornamentally for artistic signposts and other structures (C, D). On hard pines, burls or globose galls are also produced by western gall rust, but the galls caused by the rust produce orange spores in May to July. Characteristics of bark and gall tissues are also different (*see* Western Gall Rust of Hard Pines). Most burls are initiated early in the growth of the stem and they can usually be traced back to a single cell or small group of cells in the first year of shoot growth.

Cause

The cause of burls of spruce (*Picea* spp.) and pine (*Pinus* spp.) is not known; however, certain bacteria, viruses, and phytoplasma (MLO) together with insects are suspected to be the causal agents.

Distribution and Hosts

Burls are common on white and Engelmann spruces (*Picea glauca* (Moench) Voss and *P. engelmannii* Parry), especially at higher elevations in the Canadian Rockies. Similar burls are also occasionally found on lodgepole and jack pines (*Pinus contorta* Dougl. var. *latifolia* Engelm. and *P. banksiana* Lamb.).

Damage

Trees with burls seem to grow normally, but multiple main-stem burls cause inconvenience in the debarking process and produce inferior milling products.

Figure 39. Burls of spruce and pine. A. An Engelmann spruce with numerous burls. **B.** Close-up of a burl. **C.** A multiple-burled spruce log used for a signpost. **D.** Multiple-burled spruce logs used for the gate of a ranch.

A
B
C
D

Symptoms and Signs

The first external symptom of Atropellis canker is the exudation of resin from the bark surface. Resin flow increases as the canker enlarges. Sunken elongated cankers on one side of the trunk, with evidence of extensive resin flow, are the typical symptoms of this disease (A, B). Although circumference growth of cankers is very slow, they may eventually girdle and kill small trees. Large trees can be girdled only when several cankers are involved. On the center area of old cankers, small (up to 5 mm in diameter), black disk-shaped apothecia (C) and black spherical pycnidia appear, which permit a positive identification. When the fungus invades the xylem tissues of the tree, it causes blue-black discoloration of the wood and resin buildup (D).

Cause

Atropellis canker of pine is caused by a fungus, ***Atropellis piniphila*** (Weir) Lohman & Cash (Ascomycotina: Helotiales).

Distribution and Hosts

This disease is found mainly in western North America (British Columbia, Alberta, Montana, Idaho, Washington, Oregon, Arizona, and New Mexico), although it has been reported from Alabama and Tennessee. This disease is generally sporadic, but in Alberta it is widespread over most of the range of lodgepole pine (*Pinus contorta* Dougl. var. *latifolia* Engelm.) and has a high incidence in several areas. Other known hosts of this fungus are ponderosa pine (*P. ponderosa* Laws.), western white pine (*P. monticola* Dougl.), and whitebark pine (*P. albicaulis* Engelm.).

Disease Cycle

Airborne ascospores, which are produced in disk-shaped apothecia during the growing season of the host trees, are responsible for the spread of the disease. Infection occurs mainly at the branch nodes. Apothecia are produced 2–4 years after infection, and new ones are formed every year.

Damage

Atropellis canker is considered to be one of the most important diseases of lodgepole pine in Alberta. The heavy resin flow results in a debarking problem that can increase costs of processing and can degrade pulp quality. Discoloration of wood caused by the disease degrades lumber and adds to the cost of bleaching in the pulping process. The effect of this disease on tree growth is unclear, but obvious deformation of main stems must affect growth rate of infected trees. Mortality is uncommon.

Figure 40. Atropellis canker of pine, caused by *Atropellis piniphila*. A. Close-up of a canker on lodgepole pine. **B.** A stand of lodgepole pine with many infected trees. **C.** Black fruiting bodies (apothecia) of *A. piniphila*. **D.** Cross section of an infected stem showing black discoloration of wood.

A

B

C

D

Symptoms and Signs

Elongated cankers of white pine blister rust (B) girdle the stem and eventually kill the tree beyond the cankers, thus causing spiketop or flagging symptoms (A). Dome-shaped uredinia, containing urediniospores, and column-shaped telia, composed of appressed teliospores, are produced (C) on alternate hosts (*Ribes* spp.). Another rust on *Ribes* spp., ***Puccinia caricina*** DC., is often confused with ***Cronartium ribicola*** J.C. Fischer, but it produces clusters of cup-shaped aecia rather than the dome-shaped uredinia or column-shaped telia characteristic of *C. ribicola*.

Cause

White pine blister rust is caused by a rust fungus, ***Cronartium ribicola*** (Basidiomycotina: Uredinales).

Distribution and Hosts

White pine blister rust is the most destructive disease of white pines (five-needle pines) in North America, Europe, and Asia. In the prairie provinces, the hosts of this rust are whitebark pine (*Pinus albicaulis* Engelm.) (A, B) and limber pine (*P. flexilis* James) at high elevations in the Rocky Mountains, and eastern white pine (*P. strobus* L.) in eastern Manitoba. Alternate hosts are many species of gooseberry and currant belonging to the genus *Ribes*, especially wild black currant, *R. hudsonianum* Richards., bristly black currant, *R. lacustre* (Pers.) Poir. (C), and skunk currant, *R. glandulosum* Grauer (D). The rust occurs on alternate hosts beyond the range of infected pines.

Disease Cycle

Fragile, windborne spores (basidiospores) produced on alternate host plants initiate the infection on pine through the needles; then the fungus grows into the stems. Elongated cankers develop, and one to several years after infection, spermogonia with sweet droplets containing sexual spores (spermatia) are produced; later, aecia break through the bark. Powdery, orange-yellow aeciospores are released from the blisters. The aeciospores are windborne and infect alternate hosts but are incapable of reinfecting pine. Within a few weeks, small dome-shaped, spore-producing uredinia are formed, and urediniospores produced in them can reinfect alternate host leaves. Later, columnar or hairlike structures called telia are produced that consist of aggregations of teliospores. Basidiospores are produced upon germination of teliospores.

Damage

White pine blister rust was introduced from Europe to North America about 1900. It kills white pines of all ages and sizes and is considered to be one of the most destructive tree diseases in North America. Because of the scarcity of hosts, the disease is not economically important in the prairie provinces, except in eastern Manitoba, where it is found on eastern white pine. Limber and whitebark pines in the Rocky Mountains are heavily infected, but the trees are not economically important.

Figure 41. White pine blister rust, caused by *Cronartium ribicola*. A. An infected whitebark pine showing a spiketop symptom. **B.** A typical canker on young whitebark pine. **C.** Bristly black currant with uredinia and telia. **D.** A common alternate host, skunk currant.

A
B
C
D

Symptoms and Signs

The characteristic symptoms and signs of stalactiform blister rust on small stems are a slight swelling of the bark, resulting in a spindle-shaped stem, and production of orange-yellow aeciospores in early summer. On trunks of older trees, cankers are elongated, diamond-shaped, and often sunken; they spread much faster longitudinally than laterally (A, B), Cankers are made up of three distinct zones: the innermost old inactive zone, where sporulation occurred in previous years; the middle zone, where aeciospore sporulation is occurring; and the outer zone, where the fungus is advancing. On alternate host plants, uredinia and telia (D) are produced in discolored zones on the underside of leaves. The uredinia are minute, dome-shaped structures containing urediniospores; the telia are hairlike structures that produce teliospores (D).

Cause

Stalactiform blister rust of hard pines is caused by a rust fungus, ***Cronartium coleosporioides*** Arthur (Basidiomycotina: Uredinales). The name *C. coleosporioides* has been used for a group of organisms that includes western gall rust (***Endocronartium harknessii*** (J.P. Moore) Y. Hirat. (≡ *Peridermium harknessii* J.P. Moore)), limb rust (***C. filamentosum*** Hedgc.), and the fungus concerned here. However, the name should be applied only to the organism causing stalactiform blister rust. An albino form of this rust that lacks yellow pigment in the cytoplasm of all spore states is found in Banff National Park. That fungus has been named ***C. coleosporioides*** Arthur f. ***album*** Ziller.

Distribution and Hosts

This rust is found throughout Canada, from Nova Scotia to British Columbia and the Yukon, and it is particularly prominent in western Canada. Distribution extends southward into the Lake States and in the west to Colorado, Utah, and southern California. In Canada, major hosts of this rust are lodgepole pine (*Pinus contorta* Dougl. var. *latifolia* Engelm.) and jack pine (*P. banksiana* Lamb.), but it is also found on ponderosa pine (*P. ponderosa* Laws.), mugho pine (*P. mugo* Turra var. *mughus* Zenari), Scots pine (*P. sylvestris* L.), and shortleaf pine (*P. echinata* P. Mill.). Indian paint-brush (*Castilleja* spp.) and cow-wheat (*Melampyrum lineare* Desr.) are the main alternate hosts. Other known hosts for the rust are yellow owl-clover (*Orthocarpus luteus* Nutt.), lousewort (*Pedicularis bracteosa* Benth.), and yellow rattle (*Rhinanthus minor* L. (= *R. crista-galli* L.)).

Disease Cycle

Fragile, airborne basidiospores produced on the alternate host plants infect unwounded young pine shoots. Infection occurs in late summer, and cankers develop within 2 years after infection. After the production of inconspicuous spermogonia, characteristic powdery, orange-yellow aeciospores are produced. In western Canada aeciospore production continues from late May to the middle of July. Aeciospores are carried by the wind, and when they land on alternate host leaves (C) they may initiate infections. About 2 weeks after infection, uredinia appear, and when they are mature they rupture and release urediniospores. Airborne urediniospores reinfect alternate hosts, thereby repeating this state several times in one season. Later in the season, teliospores are produced on the alternate host (D), and upon germination they produce basidiospores that infect pine.

Damage

Susceptible pines of all ages are affected by stalactiform blister rust, but the greatest mortality occurs among saplings, whose main stems are killed by girdling. Isolated incidents of significant damage have been reported in young dense stands, nurseries, and tree farms in western Canada.

Figure 42. Stalactiform blister rust of hard pines, caused by *Cronartium coleosporioides*. **A.** A typical elongated diamond-shaped canker on lodgepole pine. **B.** Close-up of a canker, showing unruptured aecia. **C.** A common alternate host, Indian paint-brush. **D.** Hornlike telia on Indian paint-brush.

A
B
C
D
E
F

Symptoms and Signs

Conspicuous swelling of the bark, resulting in spindle-shaped stems, and production of orange-yellow aeciospores are the characteristic symptoms and signs of comandra blister rust on small branches. On larger stems, cankers are circular and grow laterally as much as longitudinally (A, B), thus girdling stems much faster than cankers produced by stalactiform blister rust (***Cronartium coleosporioides*** Arthur), which tends to spread much faster longitudinally than laterally. Girdling of stems above infections often causes spiketops in older stands. This is the only species of pine stem rust that has elongated, pear-shaped aeciospores with pointed tails rather than subglobose spores. Infections of this rust on nursery seedlings are often confused with infections of western gall rust (***Endocronartium harknessii*** (J.P. Moore) Y. Hirat.), because both cause similar stem swellings; however, they can be distinguished by examining whether swellings are mainly wood (western gall rust) or bark (comandra blister rust).

Cause

Comandra blister rust of hard pines is caused by a rust fungus, ***Cronartium comandrae*** Peck (Basidiomycotina: Uredinales).

Distribution and Hosts

Comandra blister rust is found across Canada, from New Brunswick to the Yukon, and southward to Tennessee, Alabama, Mississippi, New Mexico, and California. Major hosts in the prairie provinces are lodgepole pine (*Pinus contorta* Dougl. var. *latifolia* Engelm.) and jack pine (*P. banksiana* Lamb.), but it is also known on introduced pines such as ponderosa pine (*P. ponderosa* Laws.), mugho pine (*P. mugo* Turra var. *mughus* Zenari), and Scots pine (*P. sylvestris* L.). Two common alternate hosts for this disease in the prairie provinces are bastard toad-flax (*Comandra umbellata* (L.) Nutt. var. *pallida* (A. DC.) M.E. Jones) (F) and northern bastard toad-flax (*Geocaulon lividum* (Richards.) Fern.) (E). The rust often occurs on alternate hosts well beyond the range of infected pines.

Disease Cycle

Basidiospores produced on the alternate host plants are airborne and infect young pine shoots. The infection occurs in late summer, and cankers develop within 2 years after infection. After an inconspicuous sexual state (spermogonia), characteristic orange-yellow aeciospores are produced. Depending on the location, aeciospores are produced from late May to the middle of July. They are carried by the wind and infect leaves of alternate hosts. Powdery, yellow urediniospores are produced on alternate hosts and reinfect alternate hosts. This state can be repeated several times in one season. Later in the season, columns or horns of teliospores are produced on the alternate host (C–F), and upon germination they produce basidiospores.

Damage

Because of the rapid lateral growth of the cankers, comandra blister rust girdles and kills stems much faster than other species of pine stem rust such as stalactiform blister rust or sweet fern blister rust (*Cronartium comptoniae* Arthur). Bare-root nursery seedlings are often infected and killed by comandra blister rust.

Figure 43. Comandra blister rust of hard pines, caused by *Cronartium comandrae*. A. Typical circular canker on old lodgepole pine. **B.** Main stem infection on young lodgepole pine. **C.** Close-up of telial horns and dome-shaped uredinia on bastard-toad flax. **D.** Alternate states (mostly telia) on the underside of bastard toad-flax. **E.** A common alternate host, northern bastard toad-flax. **F.** Another common alternate host, bastard toad-flax.

Symptoms and Signs

Symptoms and signs of this rust on pine stems are very similar to those of stalactiform blister rust (***Cronartium coleosporioides*** Arthur), and identification of the two species is often difficult; however, the longitudinal hypertrophied ridges produced by sweet fern blister rust (A) are not produced on stems infected by stalactiform blister rust. Some minor but distinct microscopical differences also exist between the two species. This is a relatively rare disease in the prairie provinces because the range of the host plant is limited to the northern part of the region.

Cause

Sweet fern blister rust of hard pines is caused by a rust fungus, ***Cronartium comptoniae*** Arthur (Basidiomycotina: Uredinales).

Distribution and Hosts

Sweet fern blister rust is found across northern North America, from Nova Scotia to British Columbia and Alaska and south through New England to North Carolina. In Canada, major hosts of this rust are lodgepole pine (*Pinus contorta* Dougl. var. *latifolia* Engelm.) and jack pine (*P. banksiana* Lamb.). Other known hosts are shortleaf pine (*P. echinata* Mill.), mugho pine (*P. mugo* Turra var. *mughus* Zenari), Bishop pine (*P. muricata* D. Don.), ponderosa pine (*P. ponderosa* Laws.), Monterey pine (*P. radiata* D. Don.), red pine (*P. resinosa* Ait.), and Scots pine (*P. sylvestris* L.). Two alternate hosts known in nature are sweet fern (*Comptonia peregrina* (L.) Coult.) and sweet gale (*Myrica gale* L.) (B). Sweet fern only occurs in eastern Canada. Sweet gale occurs across Canada, but in the prairie provinces it is limited to northern parts of the region, thus limiting distribution of this rust.

Disease Cycle

Airborne basidiospores produced on the alternate hosts are responsible for infections on pine. Two to three years after infection, cankers and aecia containing powdery, yellow aeciospores develop. The windborne aeciospores produced on pine infect the leaves of alternate hosts (B), where uredinia develop. They contain urediniospores, which can reinfect alternate hosts. Later, columns of teliospores are produced, which germinate and release basidiospores.

Damage

This rust has caused significant damage in eastern North America in both young seedlings and larger trees, but because of the limited northern distribution of the rust, no significant economic loss has been observed in the prairie provinces. If reforestation activities and intensive management of hard pines increase in the northern region, this disease may become an important management consideration.

Figure 44. Sweet fern blister rust of hard pines, caused by *Cronartium comptoniae*. A. Infected jack pine with swollen ridges and blisters of aecia. **B.** An alternate host, sweet gale.

A
B
C
D

Symptoms and Signs

Western gall rust induces conspicuous perennial globose galls on the stems of hard pines (A–D). From May to July powdery, orange-yellow spores are produced on the surface of galls (A, D). Very young galls are sometimes spindle-shaped rather than spherical and can be confused with the spindle-shaped swellings produced by another pine stem rust, comandra blister rust (***Cronartium comandrae*** Peck). However, galls of western gall rust are mainly caused by wood (xylem) swelling, whereas galls of comandra blister rust are mainly caused by bark (phloem) swelling. This difference can be checked easily by making cross sections of small galls with a knife.

Cause

Western gall rust of hard pines (*Pinus* spp.) is caused by a rust fungus, ***Endocronartium harknessii*** (J.P. Moore) Y. Hirat. (≡ *Peridermium harknessii* J.P. Moore) (Basidiomycotina: Uredinales). Because the rust occurs across North America, several other names such as globose gall rust and pine–pine gall rust have also been proposed.

Distribution and Hosts

This rust is found across Canada, from Nova Scotia to the Yukon; in the east, southward through New York, Pennsylvania, West Virginia, and Virginia; and in the west, southward to Arizona and northern Mexico. It has often been confused with eastern gall rust (pine–oak gall rust, ***Cronartium quercuum*** (Berk.) Miyabe ex Shirai), which also produces globose galls; however, *C. quercuum* is rare in Canada and probably does not exist in the prairie provinces. In Canada the major hosts of ***E. harknessii*** are jack pine (*Pinus banksiana* Lamb.), lodgepole pine (*P. contorta* Dougl. var. *latifolia* Engelm.), and ponderosa pine (*P. ponderosa* Laws.), but it is also found on introduced pines such as mugho pine (*P. mugo* Turra var. *mughus* Zenari), Bishop pine (*P. muricata* D. Don.), Austrian or Corsican pine (*P. nigra* Arnold), maritime pine (*P. pinaster* Ait.), Monterey pine (*P. radiata* D. Don.), and Scots pine (*P. sylvestris* L.).

Disease Cycle

Unlike all other pine stem rusts in Canada, this rust has an autoecious life cycle, that is, it is capable of infecting pine directly without going to an alternate host. Spores produced on the galls (A) from the end of May to July become airborne and infect the green tissue of young shoots. Small galls appear a few months after infection but produce spores only in the year following infection. Galls grow each year and produce spores every spring for many years, unless the gall tissue dies with the stem or the causal fungus is colonized completely by mycoparasites.

Damage

Main-stem galls often kill small trees, but small active galls usually increase annually in size and produce spores each spring for many years without killing the trees. Trees with main-stem galls tend to be deformed and easily broken at the gall (B) and therefore are unsuitable for utilization. Branch galls on large trees do not significantly affect tree vigor. This disease tends to be intensified in highly managed young pine forests probably because of (1) the pine-to-pine life cycle, (2) the high susceptibility of vigorously growing shoots, and (3) the perennial nature of active galls, which serve as inoculum sources. A significant amount of western gall rust infection of nursery origin has been found in some plantations (C, D), pointing out the importance of producing disease-free planting stocks.

Figure 45. Western gall rust of hard pines, caused by *Endocronartium harknessii*. A. Two sporulating galls on a lodgepole pine. **B.** A young lodgepole pine broken at the gall. **C.** Nursery-infected seedlings of jack pine. **D.** Basal gall formation in lodgepole pine several years after planting, indicating nursery infection.

A
B
C
D

SPHAERAOPSIS (DIPLODIA) BLIGHT OF PINE AND OTHER CONIFERS

Symptoms and Signs

Tip blight and lower branch mortality are the most conspicuous symptoms of Sphaeropsis blight (A). Infected new shoots become brown and stunted with short brown needles. Infections may occur throughout the crown but damage is first evident in the lower parts of the tree. Infected dead branch tissues are resin soaked and become dark reddish brown (B, C). Stunted, straw-colored shoots with short needles are glued in place by resin. Symptoms of this disease can be distinguished from insect damage by the resin-soaked wood tissue, persistent needles, absence of insect feeding, and presence of black fruiting bodies. In the fall, black fruiting structures (pycnidia) are produced within the tissue and break through the surface of killed needles, cone scales, and twigs and branches (D). Black pycnidia on needle bases can be exposed by pulling out some of the straw-colored needles. Conidia are dark brown and mostly unicellular but occasionally two celled, 30–45 × 10–16 µm.

Cause

The causal fungus of Sphaeropsis blight is ***Sphaeropsis sapinea*** (Fr.) Dyko & Sutton (≡ *Diplodia pinea* (Desm.) Kickx) (Deuteromycotina: Coelomycetes).

Distribution and Hosts

This disease has been recorded throughout North America, but is common in central and eastern North America including Manitoba. Most pine species as well as many other conifer species are hosts of this disease. In the prairie provinces jack pine (*Pinus banksiana* Lamb.) and red pine (*P. resinosa* Ait.) are the main hosts.

Disease Cycle

Conidia are dispersed from early spring until late fall and are disseminated by splashing rain. Cone infection is important for the dissemination of the disease because the pathogen sporulates abundantly on cone scales, thus contributing to the buildup of inoculum. Wounds caused by insect feeding, hail, and other causes are important for infection of large stems.

Damage

Sphaeropsis blight can cause significant damage to jack and red pines growing under stress. On healthy unstressed trees, this disease kills only current-year buds and shoots and second-year cones, but if the trees are predisposed by such conditions as drought, root injury, and soil compaction, the pathogen also infects older twigs and branches. Extensive incidence of this disease is often recorded after successive years of drought, but it subsides once the drought condition ends. Repeated infections reduce growth and deform trees and can kill them.

Figure 46. Sphaeropsis blight of pine and other conifers, caused by *Sphaeropsis sapinea*. A. An affected jack pine. **B.** A long canker on an infected jack pine stem. **C.** Close-up of a canker on jack pine. **D.** Black pycnidia appearing between bark.

A

B

Symptoms and Signs

Typical symptoms of Scleroderris canker are orange discoloration at the base of needles in the spring, dieback of branches, and green discoloration beneath the bark of dead branches (B). Conidia are formed in black saclike pycnidia; they are falcate (new-moon-shaped) with pointed ends, four celled, hyaline, 25–35 × 24 µm. Ascospores are produced in disklike black fruiting structures (apothecia); they are ellipsoid, four celled, hyaline, 15-20 × 4 µm.

Cause

Scleroderris canker is caused by the fungus ***Gremmeniella abietina*** (Lagerb.) Morelet (= *Scleroderris lagerbergii* Gremmen) (Ascomycotina: Helotiales).

Distribution and Hosts

Scleroderris canker is known widely in Europe, Asia, and North America. Many conifer species, including pines (*Pinus* spp.) (A, B), spruces (*Picea* spp.), Douglas-fir (*Pseudotsuga menziesii* (Mirb.) Franco), and larch (*Larix* spp.), have been recorded as hosts. In the prairie provinces this disease has been discovered at a few locations in Jasper National Park on lodgepole pine (*Pinus contorta* Dougl. var. *latifolia* Engelm.) (B), and it has been found on red pine (*P. resinosa* Ait.) in western Ontario close to the Manitoba border.

Two strains have been identified in North America. The North American strain attacks young trees but causes only minor injury to trees taller than 2 m. The second strain is found in New York and Vermont as well as limited locations in Quebec and the Maritime provinces. This strain has been known to kill large pole-sized Scots pine (*P. sylvestris* L.) and red pine in northern New York and Vermont. Strain distinction is not as clear as earlier believed.

Disease Cycle

Primary infection is by windborne ascospores produced in apothecia, which appear on branches that have been dead for 1–2 years. Infections usually occur through buds or basal parts of needles; the fungus then grows into branchlets, then down to larger branches, and into the main stem of the tree. Major spore discharge and infection is in June and July. Asexual spores (conidia) ooze out of pycnidia during wet weather and are transported by rain splash to nearby branches, where they also cause infections. Pycnidia are usually produced at the base of dead needle fascicles as early as 2 months after the death of infected branchlets. Apothecia and pycnidia often occur together on dead branches.

Damage

Scleroderris canker is an economically important disease of red pine in which initial infection appears when trees are small; however, this disease is not generally a serious factor in reforestation with lodgepole or jack pines.

Figure 47. Scleroderris canker, caused by *Gremmeniella abietina*. A. Nursery infection of jack pine (*Pinus banksiana* Lamb). **B.** An infected young lodgepole pine.

A
B

LEUCOSTOMA CANKER OF SPRUCE AND OTHER CONIFERS

Symptoms and Signs

This disease is sometimes called Cytospora canker of spruce.

Cankers start on the lower branches of the tree and gradually spread upwards. Needles on infected branches turn brown and die (A). Infected needles drop off after a few months. Resin often exudes from cankers, and white or light blue patches covering the surface of bark are obvious on dead or dying branches (B). Cankers are delineated clearly, and inner bark tissue and cambium are brown in contrast to the lighter color of uninfected stems. There is usually no discoloration of xylem tissue beneath the cambium. Black fruiting bodies, pycnidia (up to 3 mm in diameter), develop in the bark; they are not visible on the bark surface but can be seen when superficial cuts are made in cankered bark. Orange spore masses (tendrils) may exude in wet weather. Spores (conidia) are one celled, hyaline, sausage-shaped, and 4–6 × 1 µm. On stems dead for several years, sexual fruiting bodies (perithecia), which are up to 0.7 mm in diameter, much smaller than pycnidia, may be present and grouped in a black stroma. Ascospores produced inside perithecia are one celled and 5–9 × 1.5 µm. The pathogen of this disease often colonizes bark killed by other causes; therefore its presence alone does not mean the fungus killed the branches.

Cause

The fungus ***Leucostoma kunzei*** (Fr.) Munk (≡ *Valsa kunzei* Fr.; stat. anamorph *Cytospora kunzei* Sacc., *Leucocytospora kunzei* (Sacc.) Urban) (Ascomycotina: Diaporthales) is the cause of this disease. Trees weakened by environmental stress are believed to be damaged by this disease, and drought is considered the most common predisposing factor.

Distribution and Hosts

This is a common disease of various species of spruce in northwestern and northeastern North America, including Quebec, Ontario, and Manitoba. Colorado blue spruce (*Picea pungens* Engelm.) sustains the greatest damage when planted outside of its natural range in the Rocky Mountains.

Disease Cycle

Conidia are released during all seasons except winter, and ascospores are released only in the spring. Both spores are spread by rain splash, wind, insects, and birds. Infection occurs through wounds.

Damage

Damage is commonly found on large trees. The disease deforms the stems and destroys the symmetry of trees. The disease itself seldom kills trees, but can cause mortality when other predisposing factors are present.

Figure 48. Leucostoma canker of spruce and other conifers, caused by *Leucostoma kunzei*. **A.** Browning of an infected branch of Colorado blue spruce. **B.** Cankered spruce stem showing bluish white resin exudate.

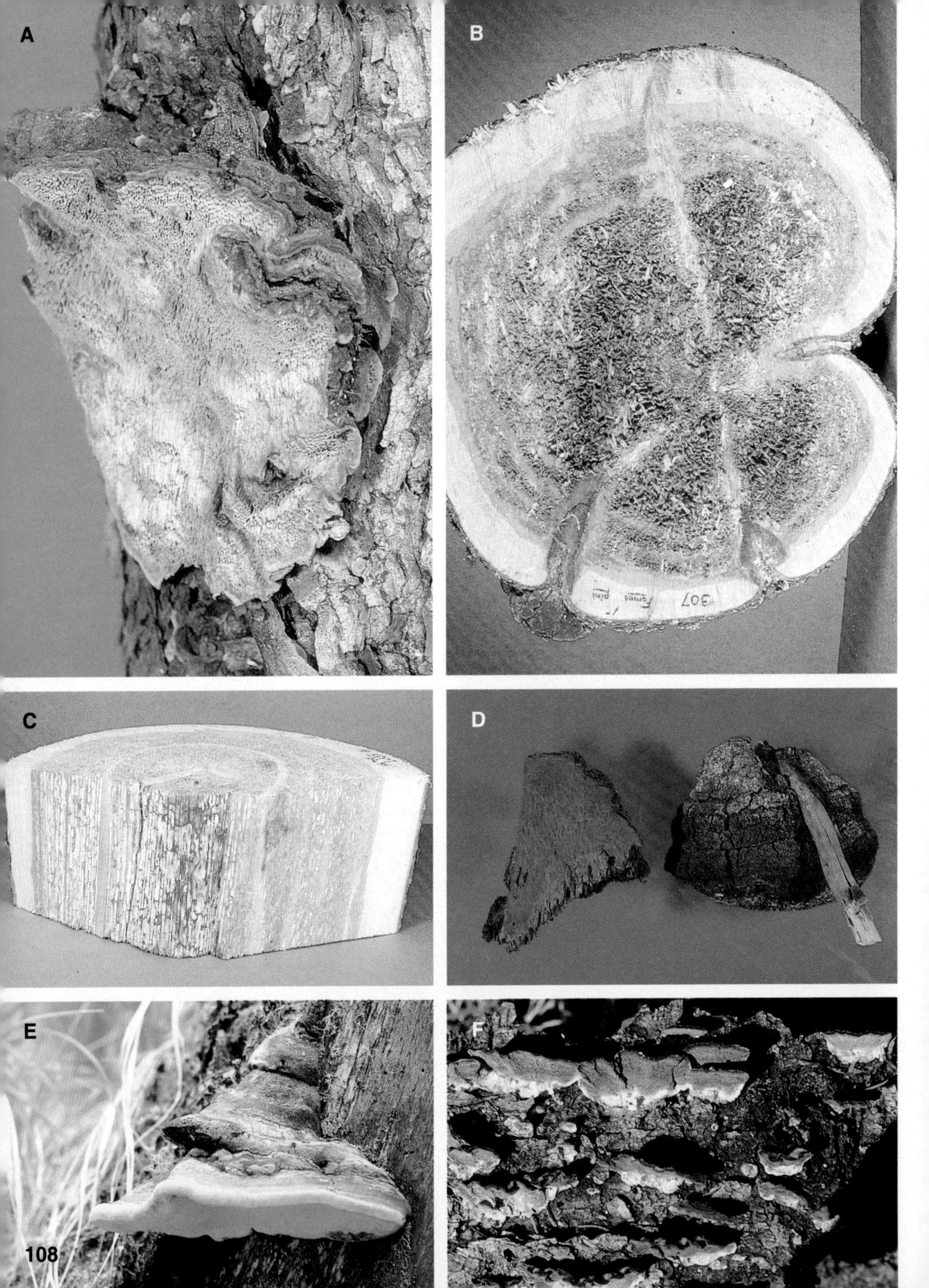
A
B
C
D
E
F

Symptoms and Signs

Identification of decay fungi is based on characteristics of decay and morphology of fruiting bodies (e.g., conks, mushrooms) associated with the decay. When fruiting bodies are not available, cultural and microscopical examinations of the fungus are necessary for positive identification.

Location of decay (heartwood, sapwood, root, butt, or trunk), type of decay (white rot or brown rot), or pattern of decay (e.g., pocket rot (50A), stringy rot (50B), or cubical rot (50C)) are different among the decay organisms (Table 4). White rot fungi are generally capable of digesting both carbohydrates (mainly cellulose and hemicellulose) and lignin, whereas brown rot fungi cannot digest lignin, and thus decayed wood appears brown.

Cause

Many fungi belonging to the subdivision Basidiomycotina decay the wood of roots, butts, and trunks of conifers. The 15 most common decay organisms associated with conifers in the prairie provinces are listed in Table 4. The red ring rot fungus (49A–C), ***Phellinus pini*** (Brot.:Fr.) Ames (≡ *Fomes pini* (Brot.:Fr.) Karst.) (Aphyllophorales), is probably the most economically important decay fungus in the prairie provinces. Armillaria root rot and Tomentosus root rot are discussed separately (*see* Armillaria Root Rot and Tomentosus Root Rot of Conifers).

Distribution and Hosts

All conifer species in the prairie provinces are attacked by various decay fungi (49A–F), and a significant amount of decay is observed, especially when trees are mature or overmature.

Disease Cycle

Airborne basidiospores produced by the fruiting bodies are the main agents of dissemination of decay fungi. Infections are initiated through branch stubs and stem wounds caused by animals, machines, and other mechanical means. Armillaria root rot (***Armillaria ostoyae*** (Romagn.) Herink) and Tomentosus root and butt rot (***Inonotus tomentosus*** (Fr.) Gilbn. and ***I. circinatus*** (Fr.) Gilbn. also spread through the forest soil or by root contacts.

Damage

Wood with advanced stages of decay cannot be used, and cull due to decay is the most important element to be considered in timber inventory and harvest planning in mature and overmature conifer stands. Trees with extensive root or butt decay have reduced physical strength and can be toppled or broken easily by high winds or heavy snow. Armillaria root rot and Tomentosus root and butt rot are known to kill young and old trees and create stand openings. Damage caused by decay fungi generally increases with the age of trees. For example, the incidence of heartrot of Engelmann spruce (*Picea engelmannii* Parry) in Alberta is 0% at 60–100 years, 17% at 101–140 years, 30% at 141–180 years, 60% at 221–260 years, and 100% at 301–340 years.

Figure 49. Decay of conifers. A–C. *Phellinus pini.* **A.** A fruiting body. **B.** Cross section of decayed lodgepole pine stem. **C.** White pocket rot of heartwood. **D.** A fruiting body of *Echinodontium tinctorium.* **E.** A fruiting body of *Fomitopsis pinicola.* **F.** Small, thin, shelving fruiting bodies of *Hirschioporus abietinus.*

Table 4. Descriptions of common decay fungi of coniferous trees in the prairie provinces

Organism	Type of decay	Fruiting body
Anisomyces odoratus (Wulf.:Fr.) Pat. ≡ *Trametes odorata* (Wulf.:Fr.) Fr.	Brown cubical pocket rot	Small annual shelving conks; upper surface velvety, reddish brown to gray; lower surface with tubes
Armillaria ostoyae (Romagn.) Herink = *A. obscura* (Pers.) Herink **Shoestring root rot, honey mushroom** (*see* Armillaria Root Rot)	Yellow to white spongy root and butt rot with numerous black zone lines; surface of decay hollow and often with dark brown wall	Honey- to brown-colored gilled mushroom with a ring (annulus) on stem (Fig. 52D)
Coniophora puteana (Schum.:Fr.) Karst.	Brown cubical rot (Fig. 50C)	Resupinate, thick, fleshy; surface olive-brown, margin cream colored
Echinodontium tinctorium (Ell. & Ev.) Ell. & Ev. ≡ *Fomes tinctorius* Ell. & Ev. **Indian paint fungus**	Brown stringy heartrot, with red streaks (Fig. 50B)	Hoof-shaped; upper surface cracked and blackish; lower surface toothed; context brick-red (Fig. 49D)
Fomitopsis officinalis (Vill.:Fr.) Bond. & Singer ≡ *Fomes officinalis* (Vill.:Fr.) Faull **Quinine conk**	Dark brown cubical rot	Large conks up to 60 cm wide, hoof-shaped, whitish
Fomitopsis pinicola (Sw.:Fr.) Karst. *Fomes pinicola* (Sw.:Fr.) Fr. **Red belt fungus**	Crumbly brown cubical rot	Large, perennial, flat, hoof-shaped conks; margin often reddish brown; upper surface crusty, gray-black (Fig. 49E)
Gloeophyllum sepiarium (Wulf.:Fr.) Karst. ≡ *Lenzites sepiaria* (Wulf.:Fr.) Fr. **Slash conk**	Brown pocket sap rot	Small, annual, shelving conks; upper surface yellow to dark brown; thin gills
Haematostereum sanguinolentum (Alb. & Schw.:Fr.) Pouzar ≡ *Stereum sanguinolentum* (Alb. & Schw.:Fr.) Fr. **Bleeding fungus, bleeding stereum**	Red heartrot	Resupinate to effused-reflexed conks; upper surface hairy, buff to gray, smooth purplish gray hymenial surface; red fluid oozes out when wounded
Hirschioporus abietinus (Pers.:Fr.) Donk ≡ *Polyporus abietinus* Pers.:Fr. **Purple conk**	White pocket rot, pitted sap rot	Small, annual, thin, shelving to resupinate conks; gray upper surface with hairs; lower surface purplish (Fig. 49F)
Inonotus tomentosus (Fr.) Gilbn. *I. circinatus* (Fr.) Gilbn. **Tomentosus root and butt rot** (*see* Tomentosus Root Rot of Conifers)	White pocket root and butt rot	Circular, stiped, hairy upper surface, yellow to rusty brown; tube layer whitish, up to 6 cm in diameter (Fig. 51A)
Peniophora pseudopini Weresub & Gib.	Pink to red stain along rays and heartwood	Small rosy-brown to brownish purple velvety crusts

Table 4. Concluded

Organism	Type of decay	Fruiting body
Phaeolus schweinitzii (Fr.) Pat. ≡ *Polyporus schweinitzii* Fr. **Velvet top fungus**	Brown cubical rot, red-brown butt rot	With or without central stalk; upper surface velvety, dark reddish brown; large angular pores
Phellinus pini (Brot.:Fr.) Ames ≡ *Fomes pini* (Brot.:Fr.) Karst. **Red ring rot**	White pocket trunk rot; affected wood becomes reddish to purplish (Figs. 49C, 50A)	Shelving, 3–20 cm wide, variable shape; upper surface dark brown to black; furrowed underside yellow-brown; brainlike irregular pores (Fig. 49A)
Pholiota alnicola (Fr.) Singer ≡ *Flammula alnicola* (Fr.) Kummer	White rot	Stiped mushroom; surface and stipe scaly
Serpula himantioides (Fr.) Karst. ≡ *Merulius himantioides* Fr.	Brown cubical rot	Resupinate patches; hymenial surface irregularly folded, brown to raw umber; margin cream colored

Figure 50. Three typical types of conifer decay. A. White pocket rot caused by *Phellinus pini* on white spruce (*Picea glauca* (Moench) Voss). **B.** Brown stringy rot caused by *Echinodontium tinctorium* on alpine fir (*Abies lasiocarpa* (Hook.) Nutt.). **C.** Brown cubical rot of lodgepole pine (*Pinus contorta* Dougl. var. *latifolia* Engelm.), probably caused by *Coniophora puteana*.

A
B
C
D

Symptoms and Signs

Major symptoms of Tomentosus root rot are excessive branch mortality from below, reduced height and increment, and general thinning of the crown. Most heavily attacked trees eventually die. Distinct openings in the stand often result and because of this, the disease is sometimes called the 'stand-opening disease' (B). Infected trees have a white pocket rot or red stain in butts and roots (C, D). Fruiting bodies of the pathogen or laboratory isolation of the fungus from diseased roots and microscopical examination are needed for positive identification. One kind of sporophore occurs on the ground. They are light brown, round, and with a short stipe (stalk) (A) (***Inonotus tomentosus*** (Fr.) Gilbn.). Another kind of sporophore is shelflike and occurs on the base of affected trees (***I. circinatus*** (Fr.) Gilbn.).

Cause

Two closely related species of fungi, ***Inonotus tomentosus*** (≡ *Onnia tomentosa* (Fr.) Karst.; ≡ *Polyporus tomentosus* Fr. var. *tomentosus*), and ***I. circinatus*** (≡ *Onnia circinata* (Fr.) Karst.; ≡ *Polyporus tomentosus* Fr. var. *circinatus* (Fr.) Sartory & Maire) (Basidiomycotina: Aphyllophorales) are the pathogens of this disease.

Distribution and Hosts

All native and introduced species of spruce (*Picea* spp.) and most species of pine (*Pinus* spp.) as well as many other conifers are the hosts of this disease. In the prairie provinces white spruce (*Picea glauca* (Moench) Voss) and black spruce (*P. mariana* (Mill.) B.S.P.) are the two main hosts. Native pines such as lodgepole pine and jack pine are less susceptible to this disease.

Disease Cycle

The main form of infection is by root contacts in which mycelium grows from a diseased root into a healthy root. Infection from dead diseased stumps can continue for at least 15 to 20 years following the death of the host. The disease likely can be disseminated over long distances by basidiospores but there is no proof of basidiospore infection.

Damage

Four types of damage have been identified: outright mortality, windthrow, growth blowdown and butt cull. Trees die either singly or in groups. If the mortality occurs over a large area it creates undesirable openings in the stand. Accumulated mortality of more than 10% of dominant and codominant trees and butt cull averaging 14% of gross merchantable volume in remaining living trees has been reported in Ontario in a white spruce plantation.

Figure 51. Tomentosus root rot of conifers caused by *Inonotus tomentosus*. A. Two fruiting bodies. **B.** Stand-opening damage caused by Tomentosus root rot. **C.** Infected root and stem of white spruce. **D.** Cross section of decayed butt of white spruce showing reddish brown stain and decay.

A
B

C

D

Symptoms and Signs

Some of the typical symptoms of Armillaria root rot are reduction in growth, yellowish green to reddish brown discoloration of foliage over the whole tree (A, B), and resinosis around the root collar. White, radiating mycelial fans formed between the bark and wood around the base of infected trees and on infected roots are typical signs (C). Dark brown or black shoestring-like fungal strands called rhizomorphs are also formed on decayed wood or in soil surrounding the diseased roots. In the fall, edible fruiting bodies (known as honey mushrooms) may be formed on diseased trees (D). Caps of the mushrooms are honey-yellow to brown, and stems have annular rings. In larger trees, the disease causes yellow stringy root and butt rot with numerous fine, black zone lines.

Cause

Armillaria root rot is one of the most important diseases of young trees in natural regeneration and in conifer plantations in the prairie provinces. The pathogen (Basidiomycotina: Agaricales), was previously considered to be a single species, ***Armillaria mellea*** (Vahl:Fr.) Kummer (≡ *Armillariella mellea* (Vahl:Fr.) Karst.), but recent studies have indicated that the disease may be caused by many closely related but distinct species of *Armillaria*. The main species causing mortality of conifers in the prairie provinces is ***A. ostoyae*** (Romagn.) Herink (= *A. obscura* (Pers.) Herink). Other species found in the prairie provinces are ***A. sinapina*** Bérubé & Dessureault and ***A. calvescens*** Bérubé & Dessureault, but they are less prevalent and economically less important than *A. ostoyae*.

Distribution and Hosts

This disease has been reported from most areas of the world and is common throughout the prairie provinces. Most species of coniferous and broadleaf trees are known to be hosts. Common and economically important tree hosts in the prairie provinces are hard pines (lodgepole pine, *Pinus contorta* Dougl. var. *latifolia* Engelm.; jack pine, *P. banksiana* Lamb.; and red pine, *P. resinosa* Ait.) and spruces (white spruce, *Picea glauca* (Moench) Voss; Engelmann spruce, *P. engelmannii* Parry; and black spruce, *P. mariana* (Mill.) B.S.P.).

Disease Cycle

The disease cycle of Armillaria root rot is not fully understood. Infection of living trees occurs mainly by rhizomorphs and by root contacts or root grafts. Rhizomorphs are usually found on diseased tissue but are known to extend up to 10 m in soil beyond the infected host plants. Spores produced on sporophores (mushrooms) disperse to dead stumps or other dead woody material and are also assumed to play some role in dissemination of the disease.

Damage

Small infected trees are usually killed quickly; large trees may have reduced growth but can keep growing for a long time in spite of the presence of the fungus in the root system or as butt rot. Often it is difficult to assess the true impact of this disease because it can kill trees already weakened by unfavorable environmental conditions, insects, and other diseases. This disease is especially important in plantations or intensively managed stands because it tends to kill young trees in groups and thus creates undesirable gaps in well-spaced stands.

Figure 52. Armillaria root rot caused by *Armillaria ostoyae*. A. An area near Hinton, Alberta, with significant mortality of regenerating lodgepole pine. **B.** Red dying lodgepole pine trees. **C.** White, fan-shaped mycelia at the base of an infected lodgepole pine. **D.** An infected lodgepole pine with young mushrooms at base.

INSECTS AND DISEASES OF CONIFERS

Seedlings

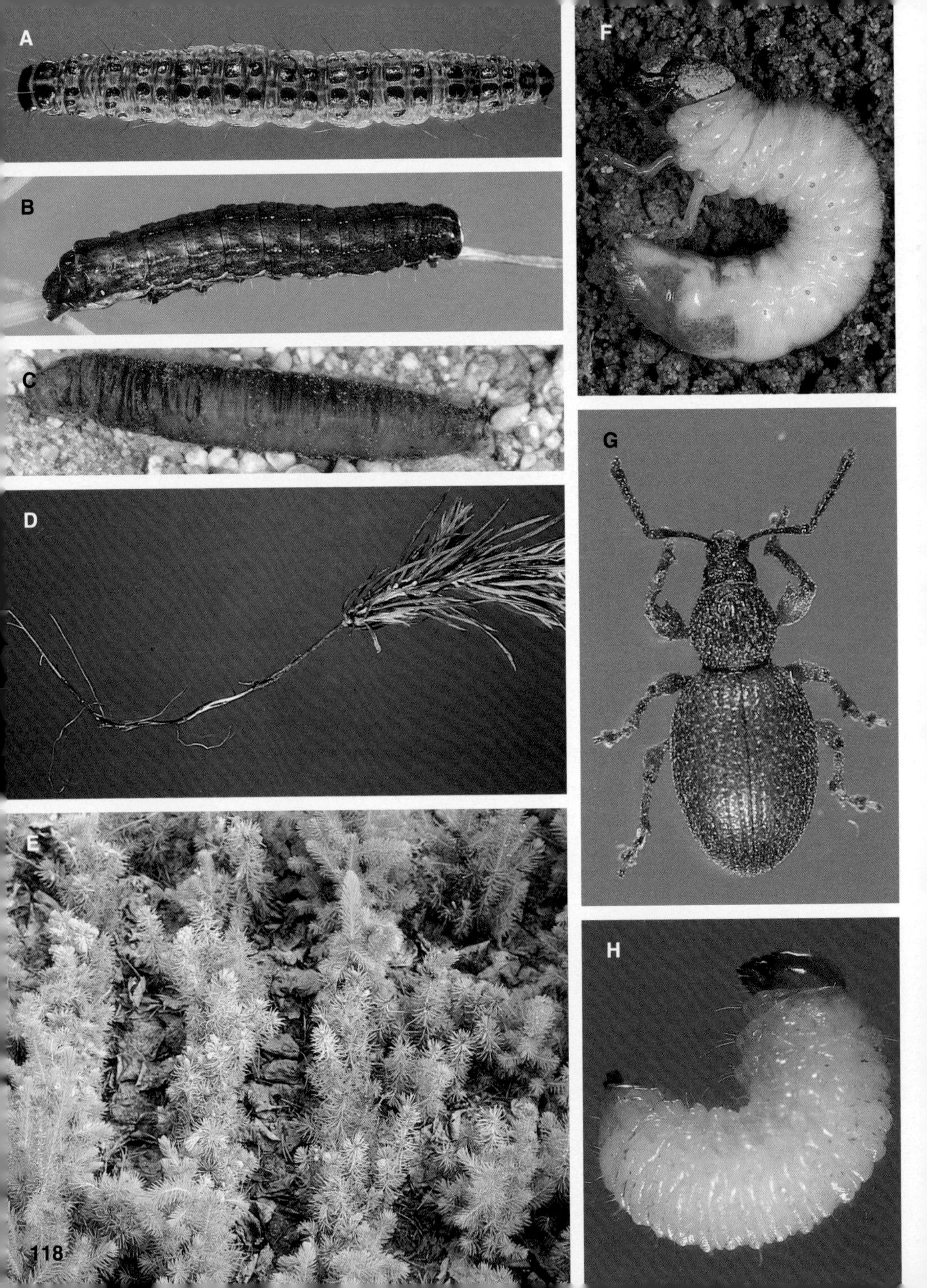
A
B
C
D
E
F
G
H

Symptoms and Signs

Only a few insects cause notable damage to conifer seedlings. Larvae of the celery stalkworm, ***Nomophila nearctica*** Munroe (Lepidoptera: Pyralidae), sever portions of seedlings above ground. Feeding of an adelgid, ***Pineus coloradensis*** (Gillette) (Homoptera: Adelgidae), causes foliage to turn yellowish (*see* Fig. 34C, Aphids, Adelgids, and Scales). Adults of the strawberry root weevil, ***Otiorhynchus ovatus*** (L.) (Coleoptera: Curculionidae), feed on foliage and tender bark above ground, and larvae sometimes sever or girdle the roots. White grub larvae, ***Phyllophaga*** species (Coleoptera: Scarabaeidae); crane flies (Diptera: Tipulidae); and larvae of a cutworm, ***Ochropleura plecta*** (L.) (Lepidoptera: Noctuidae), may also sever or girdle roots (D), killing seedlings and causing the foliage to fade (E). Identification of cause of seedling damage is difficult without observation of the damaging agent. Larvae of ***N. nearctica*** (A) have thoracic legs, five pairs of prolegs, bodies that are whitish below and purplish gray with two rows of black spots on top, and black heads and thoracic shields; they are 20–25 mm long when mature. Adult ***O. ovatus*** (G) are dark brown or black, 5–6 mm long, and wingless, and have a long snout. Larvae (H) are creamy-white with yellow-brown heads, have rudimentary thoracic legs, and are about 8 mm long at maturity. Larvae of ***O. plecta*** (B) are reddish brown with three broken dorsal white lines, have thoracic legs and five pairs of prolegs, and are 18–20 mm long when mature. Larvae of many crane flies (C) are dark in color, have no legs or head capsules, and are 15–25 mm long. ***Phyllophaga*** larvae (F) have creamy-white bodies, brown heads, thoracic legs, and no prolegs; they are 35–40 mm long when fully grown.

Distribution and Hosts

These insects are widely distributed across North America. ***Otiorhynchus ovatus*** was introduced into North America from Europe. All species usually feed on a variety of herbaceous plants but at times feed on conifer seedlings, mainly spruce (*Picea* spp.) and pine (*Pinus* spp.).

Life Cycle

Nomophila nearctica and ***O. plecta*** likely overwinter as pupae and may have two generations per year. The life cycles of many crane fly species are not well known; however, larvae predominate in late spring and early summer. ***Phyllophaga*** larvae require 3–4 years to develop. Larvae overwinter below the frost line. Adults emerge during the summer and feed on foliage of various hardwoods at night. The parthenogenetic ***O. ovatus*** has one generation per year in the prairie provinces and larvae overwinter in the soil. Adults may also overwinter in sheltered locations. Larvae pupate in the soil in the spring and new adults emerge in late spring to early summer. Adults feed on foliage and tender bark before they lay their eggs in the soil near the plants. Larvae feed on roots and mature by winter.

Damage

Insect feeding on foliage and stems may deform or kill seedlings; feeding on roots causes root deformities and often seedling death. Containerized spruce seedlings have been destroyed by ***N. nearctica*** in Manitoba and Saskatchewan. In Alberta, ***O. plecta*** has killed spruce and lodgepole pine (*Pinus contorta* Dougl. var. *latifolia* Engelm.) seedlings. Crane fly species have destroyed spruce seedlings throughout the prairie provinces. ***Phyllophaga*** species are occasionally a serious problem in red pine (*P. resinosa* Ait.) and jack pine (*P. banksiana* Lamb.) plantations in southeastern Manitoba. Lodgepole pine seedlings have been damaged by ***P. coloradensis*** in Alberta. Heavy infestations of ***O. ovatus*** adults may cause serious injury to foliage of young white cedar (*Thuja occidentalis* L.), spruce, juniper (*Juniperus* spp.), and pine; larval feeding is believed to have killed white spruce (*Picea glauca* (Moench) Voss) seedlings.

Figure 53. Insects attacking seedlings. A. *Nomophila nearctica* larva. **B.** *Ochropleura plecta* larva. **C.** Crane fly larva. **D.** Seedling girdling by a crane fly larva. **E.** Spruce seedlings killed by crane fly larvae. **F.** *Phyllophaga* larva. **G.** *Otiorhynchus ovatus* adult. **H.** *O. ovatus* larva.

A

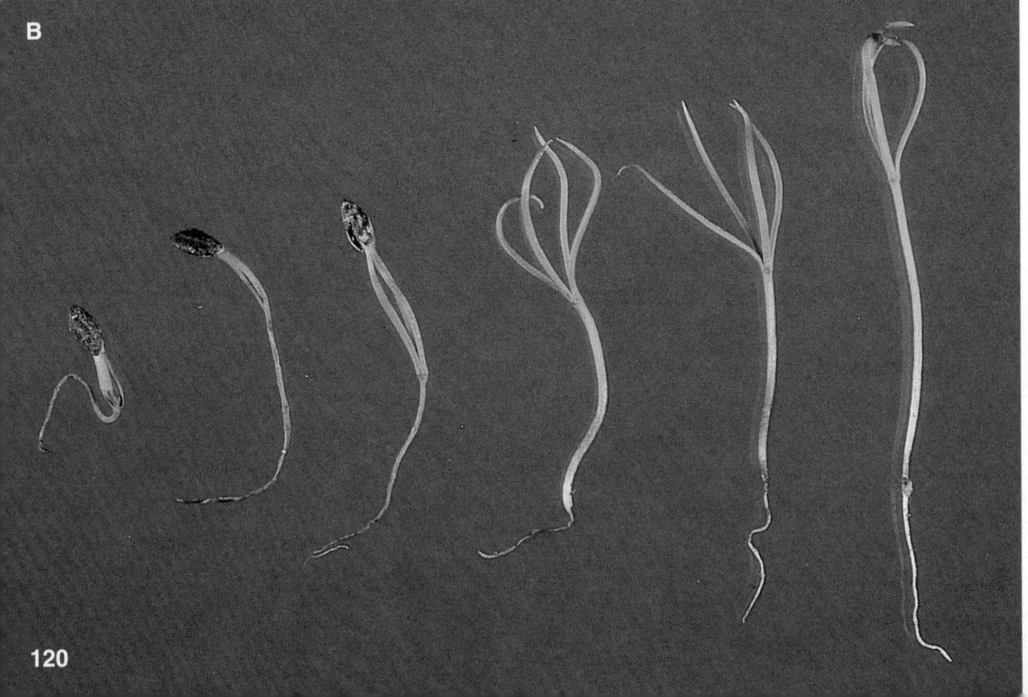
B

Symptoms and Signs

Only very young, succulent seedlings are susceptible to damping-off; once the stems of seedlings begin to develop woody tissue, susceptibility declines rapidly. A necrotic or water-soaked area develops just below the groundline and weakens the stem, causing toppling of seedlings (A, B). Color of necrotic tissue, presence or absence of visible mycelium, and speed of disease development are dependent on the fungus involved and the environmental conditions. In many situations more than one fungus may be involved. Death of viable seeds before germination or of germinated seeds before emergence is called preemergence damping-off.

Cause

Damping-off is a disease complex caused by a variety of soil-inhabiting fungi such as ***Phytophthora*** spp. and ***Pythium*** spp. (Mastigomycotina: Peronosporales), ***Fusarium*** spp. and ***Cylindrocladium*** spp. (Deuteromycotina: Hyphomycetes), and ***Rhizoctonia solani*** Kuehn (Deuteromycotina: Agonomycetes).

Distribution and Hosts

This disease complex occurs throughout the world wherever tree seedlings are grown in greenhouses and nurseries. All conifer and hardwood species are susceptible.

Disease Cycle

Most of the fungi that cause damping-off are native to nursery soil. Severity and extent of this disease are more dependent on the moisture, temperature, and nutrient status of the growing medium than simply on the presence of one or more of the pathogenic fungi in the soil. Heavy, wet, alkaline soil favors the disease, and heavy fertilization may increase incidence of damping-off. Overcrowding of seedlings, high humidity, and poor air circulation or drainage also favor damping-off.

Damage

Extensive damping-off of seedlings in nurseries is economically important. Populations of these pathogenic fungi tend to build up in nursery soils with each successive year of cropping, with average seedling losses increasing unless some control measures are implemented.

Figure 54. Damping-off of seedlings. A. Several affected seedlings of lodgepole pine (*Pinus contorta* Dougl. var. *latifolia* Engelm.). **B.** Damaged lodgepole pine seedlings showing various degrees of disease symptoms, and a healthy seedling (far right).

Symptoms and Signs

Initial evidence of molding is the presence of cottony mold (mycelium) on the lower needles of the stored seedling bundles (Fig. 55). Color and texture of fungal hyphae differ according to the species of fungus involved and age of hyphae present. They can be white, gray, orange, or various shades of brown. As the fungal growth covers the surface of seedlings, stem and needle tissues develop a water-soaked appearance and die. Affected needles often fall off. If the damage is not extensive, seedlings can be planted and the mold will usually disappear in the field.

Cause

Many kinds of fungi are associated with the molding of stored seedlings. They include genera such as ***Fusarium***, ***Aspergillus***, ***Penicillium***, ***Epicoccum***, ***Cylindrocarpon***, and ***Botrytis*** (Deuteromycotina: Hyphomycetes); ***Rhizopus*** (Zygomycotina: Mucorales); and many nonsporulating fungi, such as ***Rhizoctonia*** and others. The storage conditions of seedlings are the most important factors in the proliferation of these fungi. High relative humidity, presence of free water, high storage temperature, and long storage period are known to encourage the storage molding problem. The problem often develops rapidly after the seedlings leave controlled nursery storage facilities if suitable temperature and humidity are not maintained during transportation or storage on site. The gray mold *Botryotinia fuckeliana* (de Bary) Whet. is discussed in the next chapter (*see* Gray Mold of Stored Conifer Seedlings).

Distribution and Hosts

All tree species grown in the prairie provinces are susceptible to storage mold but they may vary in their susceptibility.

Disease Cycle

Fungi that cause storage mold often occur naturally on seedlings, soil particles, dead organic debris, and tools, and in or on containers used in nursery operations; therefore, whenever suitable environmental conditions are present during storage of seedlings, molding will occur. Spores or mycelia in contact with free water on needles and stems grow and multiply if the temperature is favorable. Growth of the fungi is superficial at first, but later results in the death of tissues. These fungi produce a large number of spores and mycelia and thus can spread quickly.

Damage

Death or damage caused by severe and extensive storage molding of conifer seedlings can cause significant economic loss, because reduction in number of healthy seedlings available for planting disrupts transportation and planting schedules.

Figure 55. Storage molds of conifer seedlings. A heavily molded bundle of white spruce (*Picea glauca* (Moench) Voss) seedlings (right) and an unaffected bundle (left).

A
B
124

GRAY MOLD OF STORED CONIFER SEEDLINGS

Symptoms and Signs

Signs and symptoms of gray mold on containerized conifer seedlings include hyphal growth on the surface of plants, discoloration of needles, wilting, and development of brownish tan, water-soaked lesions (A, B). Distinguishing characteristics are gray-brown mycelia and characteristic small, whitish spore heads with single-celled spores on the surface of affected seedlings.

Cause

Gray mold is caused by a fungus, ***Botryotinia fuckeliana*** (de Bary) Whet. (Ascomycotina: Helotiales). The fungus has been known more widely by the name of its anamorph, ***Botrytis cinerea*** Pers.

Distribution and Hosts

The gray mold pathogen is one of the most ubiquitous fungi, with a very wide host range and worldwide distribution. Although the fungus probably lives mainly as a saprophyte on plant debris, it can cause considerable economic losses in stored and transported fruits, vegetables, ornamental crops, and forest tree seedlings, especially those grown in containers.

Disease Cycle

Because this fungus is a common inhabitant of dead plant material, airborne spores (conidia) are abundant in most greenhouse or winter storage areas, especially when much dead plant debris is present under high humidity. The fungus can produce large numbers of conidia in a short time whenever proper moisture, temperature, and substrate are available.

Damage

Damage caused by this disease is becoming increasingly important as more conifer seedlings are grown and stored in containers under controlled environmental conditions. Gray mold damage can be significant, because it can affect all or a large part of a stored container-grown seedling. The problem is usually identified shortly before shipment of seedlings, and consequently plans for transportation and field planting are upset. If the damage is limited to the bottom parts of a seedling and enough healthy shoots are still available, the seedling will survive and outgrow the fungus, and the gray mold condition may eventually disappear once the seedling is planted in the field.

Figure 56. Gray mold of stored conifer seedlings, caused by *Botryotinia fuckeliana*. A. Gray mold damage on container-grown seedlings of white spruce, *Picea glauca* (Moench) Voss. **B.** Close-up of damaged white spruce seedlings.

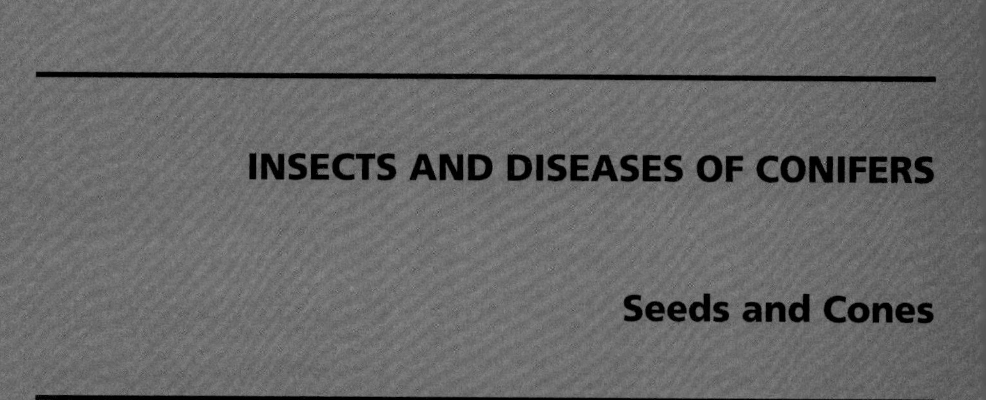

INSECTS AND DISEASES OF CONIFERS

Seeds and Cones

A

C

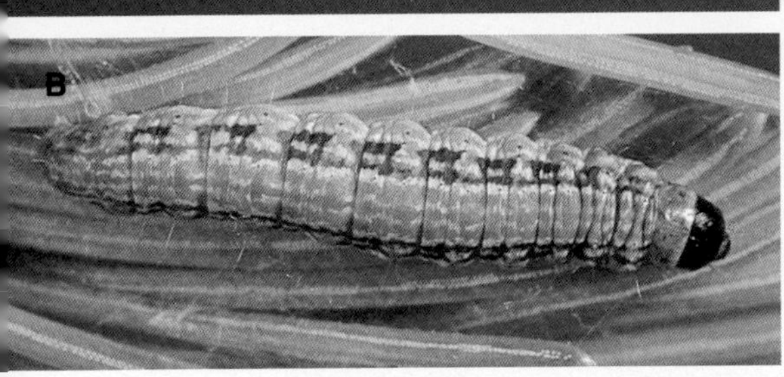
B

D

E

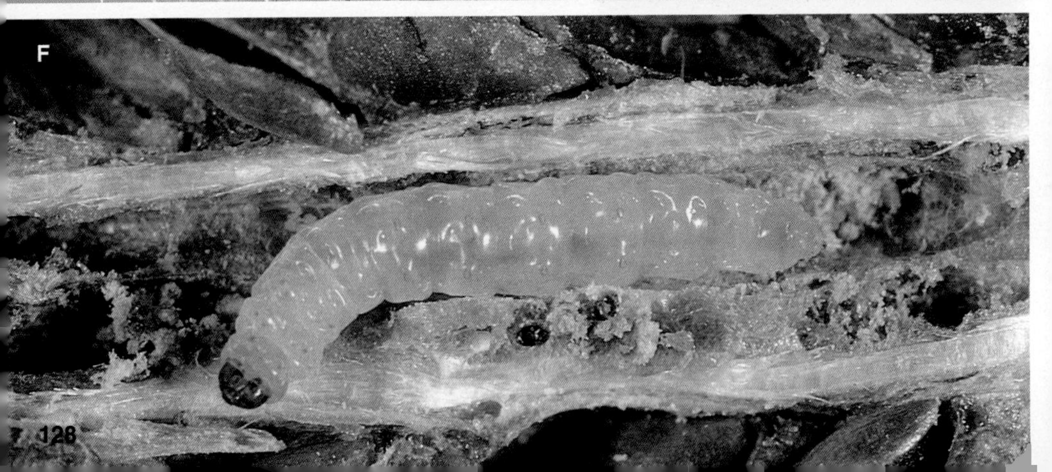
F

Symptoms and Signs

Larvae of the spruce coneworm, ***Dioryctria reniculelloides*** Mutuura & Monroe (Lepidoptera: Pyralidae), and fir coneworm, ***D. abietivorella*** (Grote), destroy most of the contents of cones. Feeding larvae expel much frass, which adheres to silken webbing and the cone's surface (D). Larvae of the spruce seed moth, ***Cydia strobilella*** (L.) (Lepidoptera: Tortricidae); and the eastern pine seedworm, ***C. toreuta*** (Grote), do not expel frass; infestation can be determined only by opening the cones to observe larvae and damage. Larvae of all four species have thoracic legs and five pairs of abdominal prolegs. Larvae of ***D. reniculelloides*** (B) have dark brown heads and pale yellow bodies with several dorsal cinnamon-colored stripes flanked by dark brown subdorsal bands, and are about 17 mm long when fully grown. ***Dioryctria abietivorella*** larvae (C) have brown heads and amber bodies but become darker as they grow and develop rows of brown spots along their backs. Fully grown larvae are about 20 mm long. Larvae of ***C. strobilella*** (F) and ***C. toreuta*** are creamy-white with dark brown heads and light brown thoracic shields, and are about 10 mm long when fully grown.

Distribution and Hosts

Cydia toreuta is distributed from Alberta to the east coast. It feeds on jack pine (*Pinus banksiana* Lamb.), red pine (*P. resinosa* Ait.), and occasionally lodgepole pine (*P. contorta* Dougl. var. *latifolia* Engelm.). The other three species have a transcontinental distribution. The ***Dioryctria*** species attack pines (*Pinus* spp.), spruces (*Picea* spp.), firs (*Abies* spp.), and Douglas-fir (*Pseudotsuga menziesii* (Mirb.) Franco). ***Cydia strobilella*** attacks spruces.

Life Cycle

Dioryctria reniculelloides has a 1-year life cycle. Young larvae overwinter in silken shelters on trees. In early spring they tunnel into needles near buds, but eventually move to feed in elongating buds or cones. Feeding is completed in late June to July, pupation occurs near feeding sites, and adults emerge after about 15 days. Eggs are laid in the tree crown. Young larvae go into hibernation without feeding.

The life cycle of ***D. abietivorella*** is variable. Some larvae pupate in the duff from July to September and adults emerge shortly afterwards to lay eggs, which overwinter on the foliage. Others overwinter as fully grown larvae in cocoons in the duff and pupate in the spring; moths emerge in spring to reproduce. Larvae feed until early fall and overwinter. Adults of both species of *Dioryctria* are similar (A); the fore wings are narrow and have a base color of brownish gray with whitish and dark gray transverse zigzag lines. The wingspan is about 25 mm.

Cydia strobilella has a 1-year life cycle. Larvae overwinter in cones, where pupation occurs in the spring. Adults emerge in late May. They are brown moths with silver bars and spots on the fore wings (E), and a wingspan of 8–11 mm. Females lay eggs between the bud scales, and the young larvae burrow towards the developing ovules. Young larvae enter seeds to feed and larger larvae chew a hole in the seed coat and feed from the outside. Damaged seeds are usually left filled with excrement. Mature larvae enter the cone axis (F) in late June and continue to feed on seeds from the axis tunnel until they hibernate. ***Cydia toreuta*** has a similar life cycle, but attacks second-year pine cones. Adults are brown, with dark markings on the fore wings.

Damage

These insects may seriously damage cones by distorting their growth and destroying seeds. Seed crops on spruce and Douglas-fir may be mostly destroyed in years when the cone crop is poor. This damage is particularly serious in seed orchards. One ***Cydia*** larva may destroy 10–20 seeds, but one ***Dioryctria*** larva may destroy an entire cone.

Figure 57. Coneworms and seed moths. A, B. *Dioryctria reniculelloides*. **A.** Adult. **B.** Larva. **C, D.** *D. abietivorella*. **C.** Larva. **D.** Spruce cone damaged by larva. Note frass and webbing. **E, F.** *Cydia strobilella*. **E.** Adult. **F.** Larva in axis of spruce cone.

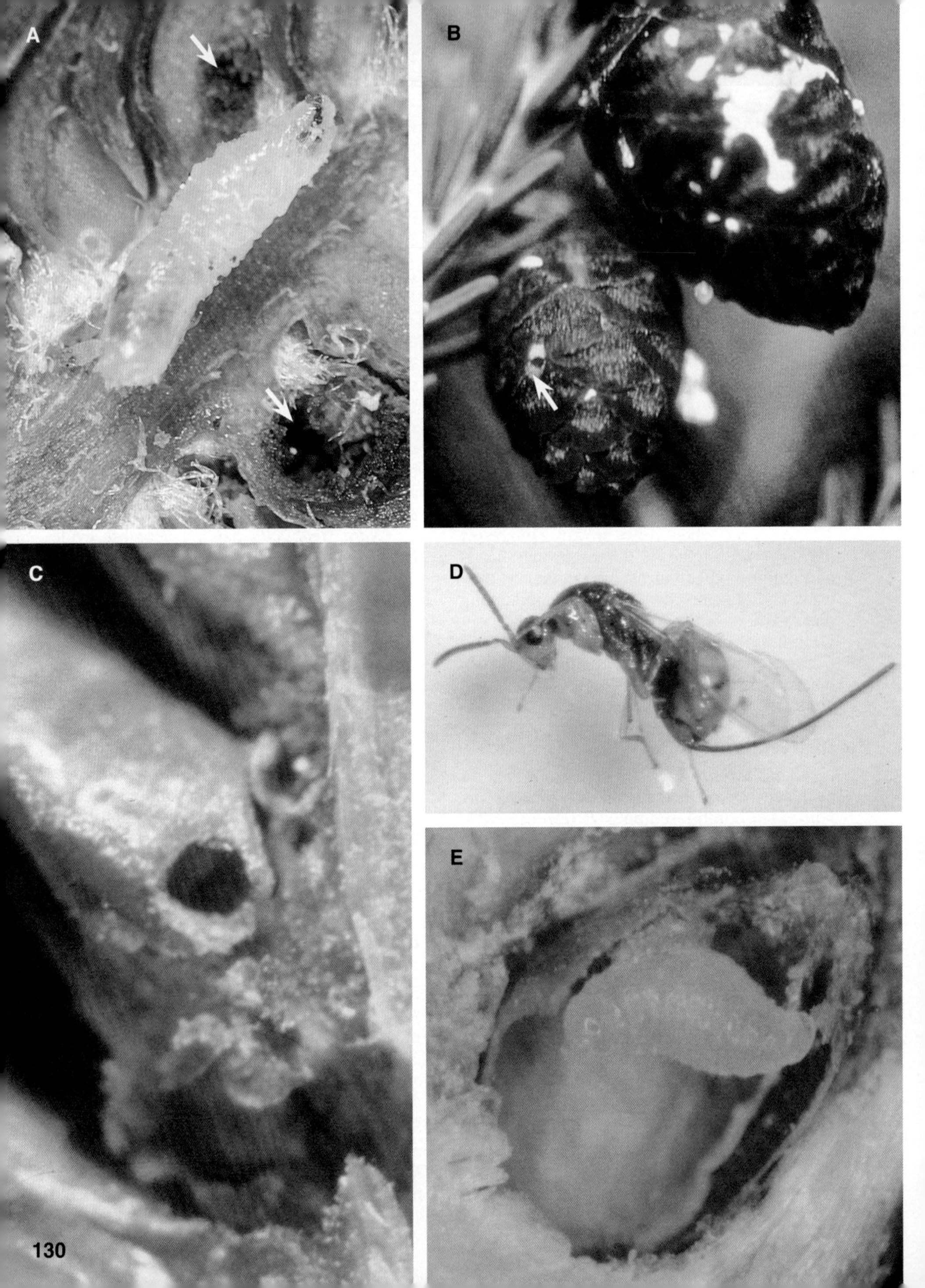
A
B
C
D
E

Symptoms and Signs

Larvae of many fly and wasp species feed in cones and seeds in the prairie provinces. The spruce cone maggot, ***Strobilomyia neanthracina*** Michelsen (Diptera: Anthomyiidae), is the most damaging maggot in spruce cones. Cones show no external signs of damage when larvae are feeding inside; however, after feeding, larvae chew circular 1- to 2-mm-diameter holes through the surface of the cone (B) and exit. Inside the cones, larvae construct a characteristic frass-filled, spiral tunnel around the cone axis and feed on the seeds. Two related species, ***S. laricis*** Michelsen and ***S. viaria*** (Huckett), cause similar damage to larch cones. Larvae of ***Strobilomyia*** (A) are whitish, have no legs or head capsule, have a distinct pair of black mouthparts, and are 5–7 mm long at maturity. The most common wasps in conifer seeds in the prairie provinces are the spruce seed chalcid, ***Megastigmus atedius*** Walker (Hymenoptera: Torymidae), and the balsam fir seed chalcid, ***M. specularis*** Walley. Larvae of both species complete their development within an individual seed. There are no external signs that cones are infested; cones must be dissected and developing seeds opened to observe larvae. Larvae are white, strongly arched, have no legs or head capsule, and are 1.5–3.0 mm long when fully grown (E). The coats of seeds infested by larvae tend to appear a little shrivelled. When adults emerge from seeds they chew a circular emergence hole, 0.5–1.0 mm in diameter, in the seed coat (C).

Distribution and Hosts

The three ***Strobilomyia*** and two ***Megastigmus*** species have boreal distributions. ***Strobilomyia neanthracina*** attacks Engelmann spruce (*Picea engelmannii* Parry) and white spruce (*P. glauca* (Moench) Voss); ***S. laricis*** and ***S. viaria*** attack tamarack (*Larix laricina* (Du Roi) K. Koch) and alpine larch (*L. lyallii* Parl.); ***M. atedius*** attacks blue spruce (*P. pungens* Engelm.), Engelmann spruce, and white spruce; and ***M. specularis*** attacks balsam fir (*Abies balsamea* (L.) Mill.).

Life Cycle

Strobilomyia neanthracina has one generation per year and overwinters as puparia in the duff. Adults emerge in early spring, just before spruce cones start to develop. Adults are 4- to 5-mm-long black flies with one pair of membranous wings. Females lay eggs between the scales of young cones during pollination. On hatching, larvae tunnel spirally around the cone axis and feed on seeds. Larvae develop through three instars. In midsummer, mature larvae chew exit holes through the cone (B), emerge, and fall to the ground to form puparia. The life cycles of ***S. laricis*** and ***S. viaria*** are thought to be similar to that of *S. neanthracina*. The different species of ***Megastigmus*** have similar life cycles. All have one generation per year, overwintering as larvae in the seeds. In the spring, adults chew small holes through the seed coats (C) and emerge. Adults have yellowish brown to blackish bodies (varies with species), are 2–4 mm long, and have two pairs of membranous wings; females have a slender, pointed ovipositor at the tip of the abdomen (D). After mating, females seek out young cones and lay their eggs inside the developing seeds. Larvae pass through five instars and complete development within the seeds.

Damage

Strobilomyia neanthracina may attack 40–100% of spruce cones in some years. One larva can destroy 50–75% of seeds in a cone and two or more usually destroy 100%. ***Megastigmus atedius*** typically attacks <10% of cones but ***M. specularis*** may occasionally cause heavy damage to cone crops.

Figure 58. Cone maggots and seed wasps. A. *Strobilomyia laricis* larva and cross section of feeding tunnel (arrows). **B.** *S. neanthracina* emergence hole (arrow) in white spruce cone. **C.** *Megastigmus atedius* exit hole in spruce seed. **D.** *M. specularis* adult. **E.** *M. atedius* larva in spruce seed.

B

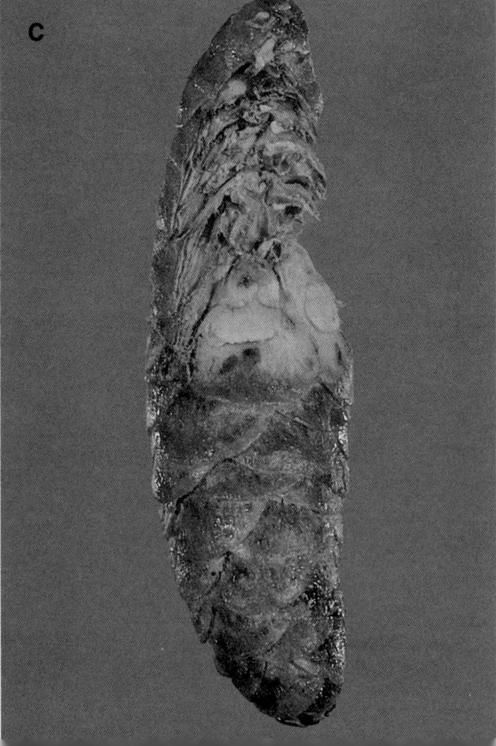
C

Symptoms and Signs

This rust attacks only cones of spruce, never the needles. The rust infects cones systemically, causing them to open prematurely. They become prematurely brown and thus can be distinguished from the healthy, closed green ones (B). Orange-yellow aeciospores are produced on diseased cones (B, C). In certain years, spores from infected cones are produced in such quantity that the forest floor or lake surface appears orange because of the fallen spores.

Several species of spruce needle rust, such as ***Chrysomyxa ledicola*** Lagerh. and ***Pucciniastrum americanum*** (Farlow) Arthur, occasionally infect cones, but unlike ***C. pirolata*** Wint., their infections are localized and not systemic. On alternate hosts, uredinia and telia are produced on the lower surface of the leaves. The infected leaves are often smaller and yellowish. Pustules of uredinia are powdery and orange (A). Telia develop as flat, waxy, yellowish red pustules.

Cause

Spruce cone rust is caused by a rust fungus, ***Chrysomyxa pirolata*** Wint. (Basidiomycotina: Uredinales).

Distribution and Hosts

Spruce cone rust is distributed in patches across northern North America, Europe, and Asia. All native North American spruce are susceptible, but white spruce (*Picea glauca* (Moench) Voss), black spruce (*P. mariana* (Mill.) B.S.P.), and Engelmann spruce (*P. engelmannii* Parry) are natural hosts in the prairie provinces. Introduced blue spruce (*P. pungens* Engelm.) is also reported to be susceptible. Alternate hosts of this fungus are several species of wintergreens (*Orthilia* spp., *Pyrola* spp., and *Moneses* spp.). The main alternate hosts in the prairie provinces are one-sided wintergreen (*Orthilia secunda* (L.) House ≡ *Pyrola secunda* L.), common pink wintergreen (*Pyrola asarifolia* Michx.; A), and one-flowered wintergreen (*Moneses uniflora* (L.) A. Gray).

Disease Cycle

Basidiospores produced on alternate hosts infect female flowers of spruce, probably during or soon after pollination. Systemically infected cones produce aeciospores, which are responsible for infection on alternate hosts. On alternate hosts, uredinia and telia are produced on systemically infected leaves (A). Urediniospores spread the rust to other alternate hosts. Teliospores germinate to produce basidiospores.

Damage

Most of the infected cones do not produce viable seeds. Up to 50% of the cone crop can be infected, and in certain locations damage to 20–30% of the cone crop is common. In selected harvesting areas or seed orchards, high levels of infection during cone crop years are economically important; however, healthy seeds superficially contaminated with the rust spores from infected cones cannot be infected.

Figure 59. Spruce cone rust caused by *Chrysomyxa pirolata*. A. Systemically infected (left) and healthy (right) common pink wintergreen. **B.** An infected cone (brown open cone) and two uninfected green cones of white spruce. **C.** An infected cone of white spruce cut open to show yellow systemically produced aecia.

INSECTS AND DISEASES OF BROADLEAF TREES

Foliage and Buds

A
B
C
D
E
F
G

Symptoms and Signs

Grayish brown egg bands of the forest tent caterpillar, ***Malacosoma disstria*** Hübner (Lepidoptera: Lasiocampidae), on twigs (B, C) from late summer to early spring indicate the presence of this species and herald forthcoming defoliation. Larvae hatch in late April to early May. They are black, hairy, and 2–3 mm long. Larvae feed gregariously (E), initially on opening buds, later on leaves. Larvae do not roll or tie leaves together. Defoliation progresses from the outer tree crown inward and downward, and is usually complete by mid-June. When not feeding, larvae rest in a mass on a silken mat spun on the trunk or large branches. If all foliage is consumed before larvae complete development, they migrate and feed on understory shrubs and other vegetation. Mature larvae (D) are 45–55 mm long, hairy, and have three pairs of thoracic legs and five pairs of abdominal prolegs. They have broad, bluish lateral bands and narrow, broken, orange and brown lines on the body, and white or creamy-white keyhole-shaped dorsal markings. Silken cocoons are usually spun between aspen leaves (F), but may occur on other sites if trees are completely defoliated. Several other species and subspecies of ***Malacosoma*** occur on shrubs and small saplings in the prairie provinces; however, larvae of these species lack the keyhole-shaped dorsal spots.

Distribution and Hosts

This species has a transcontinental distribution and feeds mainly on trembling aspen (*Populus tremuloides* Michx.). During severe infestations, larvae readily feed on many other hardwood tree and shrub species.

Life Cycle

The forest tent caterpillar has one generation per year. Larvae hatch in late April and early May, about the time that aspen buds flush. They develop through five instars and complete development by about mid-June. Mature larvae spin cocoons, usually between aspen leaves (F), and pupate. Moths emerge about 10 days later. Moths (A) are yellowish brown, stout, and hairy; the fore wings have two oblique dark bands, and the wingspan is 35–45 mm. Mated females deposit eggs in bands around small twigs and cover them with frothy spumaline (B, C). The embryos become fully developed larvae in about a month, but larvae do not emerge until the following spring.

Damage

Early larval feeding may kill aspen buds. One or more years of severe defoliation (G) may result in twig mortality, reduced radial growth of the stem, and smaller leaf size. Aspen trees are sometimes killed by several years of severe defoliation, but this usually occurs in conjunction with other stress factors, such as summer drought or late-spring frosts. The risk of aspen mortality is minimized because trees refoliate 3–6 weeks after defoliation. Defoliated trees are usually more susceptible to stem cankers, decay, and bark- and wood-boring insects. Radial stem growth on severely defoliated trees may be only 10% of normal. In June, masses of migrating caterpillars may be a nuisance in recreational and public areas. Cocoons on buildings or ornamental vegetation may be unsightly and require removal. In July, large numbers of adult moths attracted to outdoor lights are a nuisance and also may result in localized infestations.

Figure 60. Forest tent caterpillar, *Malacosoma disstria*. A. Adult. **B.** Egg mass covered with spumaline. **C.** Egg mass with some spumaline removed to show eggs. **D.** Mature larva. **E.** Colony of larvae feeding on aspen leaves. **F.** Cocoon spun among aspen leaves. **G.** Severely defoliated aspen stand.

A
B
C
D
E
F
G

Symptoms and Signs

Young larvae of the large aspen tortrix, ***Choristoneura conflictana*** (Walker) (Lepidoptera: Tortricidae), begin mining buds in early spring; however, damage caused at this time is usually inconspicuous. Later larval instars continue to feed within rolled leaves or within two or more leaves pulled together and secured with silken webbing (F). Thus, affected foliage has a clumped, irregular appearance and leaves do not move as freely in the wind as uninfested leaves. Larvae have three pairs of thoracic legs and five pairs of abdominal prolegs. Young larvae (B) are 3–4 mm long, have black heads, and yellowish or pale green bodies that become progressively darker with each molt. Mature larvae (D) are 15–21 mm long, are dark green with black heads, thoracic shields, and anal patches, and have two rows of small paired black spots along the back. When population densities are high, silk produced by dispersing larvae is conspicuous on trees and understorey vegetation (G). In mid-June, pupation occurs within the webbed leaves. Empty blackish pupal cases are often seen protruding from rolled leaves or leaf clusters (E). Eggs are laid in clusters on the upper surface of leaves in July; they are light green and appear as overlapping scales (B). Young larvae construct silken hibernacula (C) underneath bark scales, dead bark, or moss on tree stems and branches. The large aspen tortrix can be differentiated from other caterpillar species that roll or web leaves by the appearance of the larvae, pupation site, color of pupae, presence of egg masses on foliage, and by the presence of hibernacula.

Distribution and Hosts

This species is distributed from coast to coast, as far north as southern parts of the Yukon and Northwest Territories, and in adjacent parts of the United States. It feeds mainly on trembling aspen (*Populus tremuloides* Michx.) but also feeds on other hardwood species when population densities are high.

Life Cycle

The large aspen tortrix has one generation per year. Overwintering second-instar larvae emerge from hibernacula (C) when aspen buds begin to swell (about 10 days before leaves appear). Larvae mine the buds, where they molt once. Late instars feed within leaves rolled or webbed together. Pupation occurs on the foliage in mid-June; adults emerge about 10 days later. Adult moths (A) have a wingspan of 25–35 mm, are brownish gray, and the fore wings are light gray with three patches of darker gray. Mated females lay eggs in clusters of 50–450 on the upper surface of leaves (B). Eggs hatch in about 10 days and larvae feed gregariously on epidermal leaf tissue between two leaves webbed together with silk. This feeding injury is relatively insignificant. In mid-August the young larvae cease feeding, seek out suitable hibernation sites, and spin hibernacula, in which they overwinter.

Damage

Severe defoliation causes radial growth loss, but little tree mortality. Several successive years of moderate to severe defoliation by *C. conflictana* and other co-occurring defoliators on aspen may stress trees to the point where they become more susceptible to stem cankers, decay, and bark- and wood-boring insects. In June, masses of migrating caterpillars and their associated webbing are an annoyance in recreational areas and campgrounds.

Figure 61. Large aspen tortrix, *Choristoneura conflictana*. A. Adult. **B.** Egg mass and newly hatched larvae. **C.** Overwintering larvae in hibernacula. Webbing is removed from the top of one hibernaculum to expose a larva. **D.** Mature larva. **E.** Empty pupal case protruding from cluster of webbed aspen leaves. **F.** Groups of aspen leaves webbed together by feeding larvae. **G.** Severely defoliated aspen stand. Note the large amounts of silk spun by wandering larvae.

A
B
C
D
E
F

G

Symptoms and Signs

Larvae of the Bruce spanworm, ***Operophtera bruceata*** (Hulst) (Lepidoptera: Geometridae), begin feeding by mining developing aspen buds in the spring. This damage is initially inconspicuous, but when leaves expand, this damage appears as holes in the leaves. As leaves expand, larvae may roll or web leaves together and feed within the enclosed space, or they may feed openly, giving the foliage a ragged appearance (F). If they destroy the entire leaf crop, larvae often leave the trees on silken strands. The trees and understory vegetation then appear to be covered in a silken shroud (G). Larvae (D, E) are stout-bodied loopers, with three pairs of thoracic legs and two pairs of abdominal prolegs, and are about 18 mm long when fully grown. They are typically light green and have one prominent and two indistinct yellowish lines along each side of the body and dark brown heads (E). There is a large amount of color variation, however, and some individuals have blackish heads and dark gray bodies with three whitish bands or broad stripes along each side (D). In the absence of larvae, damage caused by Bruce spanworm may be confused with that caused by large aspen tortrix (*see* Large Aspen Tortrix); however, if the damage is caused by Bruce spanworm, no pupal cases or egg masses will be present on the foliage, and hibernacula will not be present on nearby stems and branches.

Distribution and Hosts

This species is distributed from coast to coast in Canada and from the Lake States to New England in the United States. Its principal host in the prairie provinces is trembling aspen (*Populus tremuloides* Michx.), but it also feeds on many other hardwood tree and shrub species.

Life Cycle

The Bruce spanworm has one generation per year and overwinters in the egg stage. Eggs are deposited in bark crevices or in moss at the base of tree trunks. They are pale green at first and turn bright orange (C). Larvae hatch in the spring near the time when aspen foliage flushes. Larvae pass through four instars and complete development in 5–7 weeks, in mid- to late June. Mature larvae drop to the leaf litter and spin light silken cocoons in the duff, in which they pupate. In late fall, once frosts have occurred, adults emerge from cocoons in the late afternoon or early evening of warm days. Female moths (B) are wingless, covered in rough scales, and are brownish in color. Male moths (A) have light brown slender bodies, light brown and gray fore wings with dark gray bands, and wingspans of 25–30 mm. Females climb the tree trunks and release a pheromone that attracts males. Males can fly at low temperatures. After mating, females lay eggs and die.

Damage

Several outbreaks of the Bruce spanworm on aspen have been recorded in Alberta since 1903. Recent outbreaks have occurred toward the end of each of the last four decades. At its peak, the outbreak in 1958 covered 130 000 ha of aspen forest. In general, severe defoliation of aspen forests lasts 2–3 years and results in reduced radial growth of trees. Bruce spanworm outbreaks may coincide with, be preceded by, or followed by, outbreaks of other aspen defoliators. Thus, Bruce spanworm may contribute to tree mortality if severe defoliation occurs for several consecutive years. In June, dispersing larvae and associated silk (G) webbing are an annoyance in recreational areas and campgrounds.

Figure 62. Bruce spanworm, *Operophtera bruceata*. A. Adult male. **B.** Adult female. **C.** Eggs in aspen bark crevices. **D.** Mature larva, dark color form. **E.** Mature larva, light color form. **F.** Close-up of larval feeding damage on aspen foliage. **G.** Aspen stand severely defoliated by Bruce spanworm. Note large amounts of silk spun by wandering larvae.

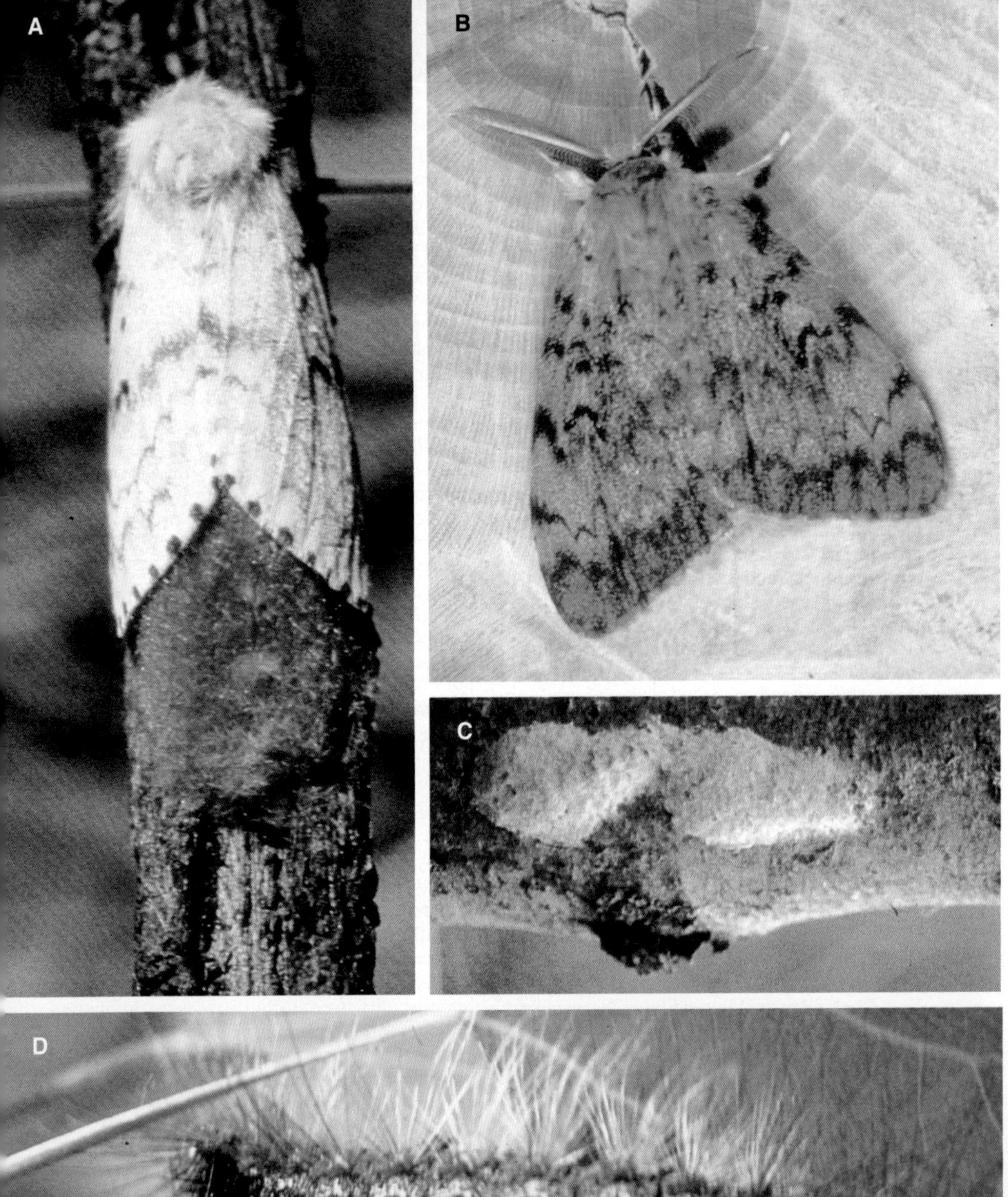
A
B
C
D

Symptoms and Signs

In eastern North America, larvae of the gypsy moth, ***Lymantria dispar*** (L.) (Lepidoptera: Lymantriidae), begin feeding on foliage in late April to mid-May. The first three larval instars alternate between feeding and resting during the day. First-instar larvae chew small holes in the leaves and rest on a mat of silken threads on the underside of the leaves. Second and third instars feed at the leaf margins and rest beneath branches and twigs. Older larvae feed at night and descend from the trees at dawn to rest in protected locations. About 85% of the damage to leaves is done by the last larval instar in late June and early July. Young caterpillars are dark and hairy. In older larvae, the yellowish head and yellowish brown, hairy body are both densely mottled with black; five pairs of dorsal blue spots occur on the anterior portion of the body and six pairs of red spots more posteriorly (D). Mature larvae are 35–60 mm long. All instars have thoracic legs and five pairs of prolegs. Pupae are reddish brown, usually occur in clumps, and are not enclosed by cocoons. Eggs occur in masses that are covered by a mat of yellowish brown hairs from the female's body (A). Each mass may measure 25–40 mm at its widest point. Eggs are deposited on the underside of branches, on tree trunks (C), under bark scales, or in other sheltered places, where they remain from July to April.

Distribution and Hosts

The gypsy moth was introduced from Europe into eastern North America in the 19th century and is spreading westward. This species is also present in British Columbia and several western states. Over the last 10 years, several male moths have been trapped in pheromone traps in the prairie provinces. In 1993, an egg mass was discovered in Red Deer, Alberta, on a military cargo shipped from Germany. In the spring of 1994, a colony of larvae hatched from an egg mass on an imported blue spruce (*Picea pungens* Engelm.) planted near Steinbach, Manitoba; these were destroyed. The gypsy moth has not yet become established in the prairie provinces. Larvae feed on over 500 species of plants worldwide. In North America, oaks (*Quercus* spp.), trembling aspen (*Populus tremuloides* Michx.), and willows (*Salix* spp.) are among the most susceptible species. When severe infestations occur, larvae feed on almost any species of tree or shrub, including conifers. In the early 1990s, an Asian form of the gypsy moth was detected in the Vancouver area of British Columbia. It is believed that this infestation was eradicated.

Life Cycle

In eastern North America, this species has one generation per year and overwinters in the egg stage. Eggs hatch from late April to mid-May. First-instar larvae are easily dispersed by the wind. Larvae complete development by early July and pupate. Moths emerge 10–14 days later. Adults of both sexes have wings but only males can fly. Males have brown bodies, light to dark brown wings with black markings, and a wingspan of 28–35 mm (B). Females have a wingspan of 40–60 mm and are mainly white, but the abdomen is covered in yellowish brown hairs. The fore wings have several blackish oblique bands and a transverse line of black dots near the outer border (A). Females of the European form of gypsy moth do not fly; however, females of the Asian form are capable of flight. After mating, females lay eggs in late July and August.

Damage

The effects of gypsy moth defoliation vary and depend on the condition of trees before defoliation, the amount of foliage removed, and the number of years of consecutive defoliation. Radial growth of defoliated trees may be reduced up to 50%. Healthy trees can withstand one or two consecutive defoliations, but stressed trees may die after two or three consecutive defoliations. Weakened trees are often attacked and killed by other opportunistic insects and fungi. Also, the recreational, aesthetic, and watershed values of forest, park, and ornamental trees can be seriously threatened by this pest. The masses of caterpillars existing during outbreaks are also a public nuisance.

Figure 63. Gypsy moth, *Lymantria dispar*. A. Female moth laying eggs. **B.** Male moth. **C.** Overwintering egg masses on trunk of tree. **D.** Mature larva.

A
B
C
D
E
F
CN

Symptoms and Signs

Overwintered larvae of the satin moth, ***Leucoma salicis*** (L.) (Lepidoptera: Lymantriidae), begin feeding in mid-May. They feed openly and consume whole leaves, except the major veins. Damage is most conspicuous after mid-June, when larvae are large and consume large quantities of foliage. Larvae (D) have thoracic legs and five pairs of abdominal prolegs, and are 35–45 mm long when fully grown. The basic body color is grayish brown, and the head and back are dark. There is one row of large, oblong white or pale yellow patches along the middle of the dorsal surface and two subdorsal yellowish lines. The two lateral and two subdorsal rows of orange tubercles have tufts of long brownish setae attached to them. Larvae molt on the underside of branches, and cast skins are conspicuous there. Pupae (E) are shiny black, 15–22 mm long, and have tufts of yellowish hairs. They occur in loosely woven silken cocoons in rolled leaves, on twigs, or in bark crevices. If it is too late in the season to observe live mature larvae and pupae, the presence of rolled leaves containing pupal cases and of empty larval skins and silken webbing on boles and branches indicate satin moth infestations.

Distribution and Hosts

The satin moth was introduced into North America from Europe. It was first detected near Boston, Massachusetts, and in southeastern British Columbia in 1920. In the east, the species is currently distributed from Newfoundland, through eastern Canada and the northeastern United States, to Ontario. In the west, it is distributed from British Columbia to northern California. In 1994, an infestation of the satin moth was detected in Edmonton and St. Albert. The species has apparently been present in Edmonton since 1991 and is well established. Satin moths feed on all species of poplar (*Populus* spp.) and willow (*Salix* spp.), but prefer ornamental poplars. They also occasionally feed on other hardwood species.

Life Cycle

This species has one generation per year. Third-instar larvae overwinter in silken, cocoonlike hibernacula that are usually covered with bark particles, mosses, or lichens, and attached to the trunks or branches of host trees. Larvae emerge in mid-May and feed on leaves until late June or early July. There are seven or eight larval instars. Mature larvae spin silken cocoons in the leaves, in which they pupate (E). Adults emerge throughout most of July. Adults (A) are satin-white moths with black, robust bodies that show through the hairs. There are no markings on the wings and the wingspan is 24–47 mm. Mated females lay eggs in clusters on leaves, branches, or trunks. Eggs are light green, flat, and laid in oval masses of 150–200 eggs covered with a glistening, white secretion (B). Eggs hatch after about 2 weeks and young larvae (C) move to the leaves, which they skeletonize as they develop through two instars. Second-instar larvae seek out hibernation sites, spin silken hibernacula, and molt to third instars, which overwinter.

Damage

Satin moths are capable of completely defoliating trees (F). Trees often refoliate after severe defoliation, but the second crop of leaves is smaller and lighter in color. Severe defoliation in consecutive years results in reduced radial growth of the stems, branch mortality, and sometimes tree mortality. The impact of defoliation may be more severe on trees already stressed by other factors such as drought. Weakened trees may be attacked by other opportunistic insects and fungi. Larvae migrating from completely defoliated trees to new food sources may also be a public nuisance.

Figure 64. Satin moth, *Leucoma salicis*. A. Adult. **B.** Egg mass. **C.** Young larva and skeletonized poplar leaf. **D.** Mature larva. **E.** Pupae in silken cocoons. **F.** Poplar tree severely defoliated by satin moth larvae.

A
B
C
D
E
F
G

Symptoms and Signs

Larvae of the uglynest caterpillar, ***Archips cerasivorana*** (Fitch) (Lepidoptera: Tortricidae), and the fall webworm, ***Hyphantria cunea*** (Drury) (Lepidoptera: Arctiidae), construct silken tents (F, G) on small hardwood trees and shrubs. ***Archips cerasivorana*** nests are the more common. It is difficult to differentiate between the two species based on tent architecture. Larvae are gregarious and live inside the tents. Fully grown ***A. cerasivorana*** larvae are 20–23 mm long, and have yellowish green bodies with small dark spots and dark heads, thoracic shields, and anal shields (D). Larvae of ***H. cunea*** (E) have pale yellowish bodies with a broad black dorsal band. Tubercles on the band are also black, but those on the sides are orange. Long silky hairs arise in tufts from the tubercles. Mature larvae are about 25 mm long. If larvae are absent, identifications may be based on cast skins of larvae or the presence or absence of pupal cases; ***A. cerasivorana*** pupates within the nests but ***H. cunea*** does not. Similar tents are also constructed by the oak webworm, ***A. fervidana*** (Clemens), on bur oak (*Quercus macrocarpa* Michx.) in Manitoba; however, this species is uncommon. ***Malacosoma*** (Lepidoptera: Lasiocampidae) nests occur on tree branches and stems and rarely enclose foliage, and larvae occur on the surface.

Distribution and Hosts

Archips cerasivorana and ***H. cunea*** occur from coast to coast in Canada and the northern United States. Choke cherry (*Prunus virginiana* L.) is the main host of ***A. cerasivorana***, and ***H. cunea*** attacks mainly birch (*Betula* spp.), cherry (*Prunus* spp.), willow (*Salix* spp.), and elm (*Ulmus* spp.). Both species may attack many other hardwood tree and shrub species.

Life Cycle

Archips cerasivorana has one generation per year and overwinters as eggs, which occur in dark brownish, flattened masses (B), usually on the main stem near the base of shrubs. Larvae hatch as soon as choke cherry leaves flush and crawl to the top of shrubs, where they begin webbing leaves together to form protective nests. Larvae feed on the foliage enclosed by the nest (F). Nests are enlarged as necessary to accommodate the food requirements of the colony. Mature larvae pupate in silken cells inside the nest. Pupae wriggle out of their cells and move to the outside of the nests, where the adults emerge. Empty pupal cases remain attached to the nest. Adults (A) are dull yellowish orange and have wingspans of 20–24 mm. The fore wings are crossed with orange and silvery-white striations and have several dark brownish patches. Adults are active from late June to early September, when egg laying occurs.

Hyphantria cunea has one generation per year in the prairie provinces and overwinters in the pupal stage. Adults emerge from late June to mid-July. They have a wingspan of 30–40 mm and are white without marks; however, the yellowish bodies often show through the white hairs on the abdomen (C). After mating, females lay eggs in masses of up to 300 on the underside of leaves. Eggs hatch in about 2 weeks and young larvae web leaves together to construct a nest. Larvae feed throughout the summer on foliage enclosed by the nests (G). Mature larvae leave the nests to seek sheltered spots for pupation. Cocoons are woven in bark crevices, beneath stones, or in the duff, and pupae overwinter.

Damage

Both species can be a problem in nurseries and ornamental plantings because of the unsightly appearance of the nests. Nests may also cause some branch deformity in subsequent years, but little other permanent damage. In the forest, most nests are on undergrowth and are seldom abundant enough to be of major concern.

Figure 65. Uglynest caterpillar and fall webworm. A. Adult *A. cerasivorana*. **B.** Egg mass of *A. cerasivorana*. **C.** Adult *H. cunea*. **D.** Mature *A. cerasivorana* larva. **E.** Mature *H. cunea* larva. **F.** *A. cerasivorana* nest. **G.** *H. cunea* nest.

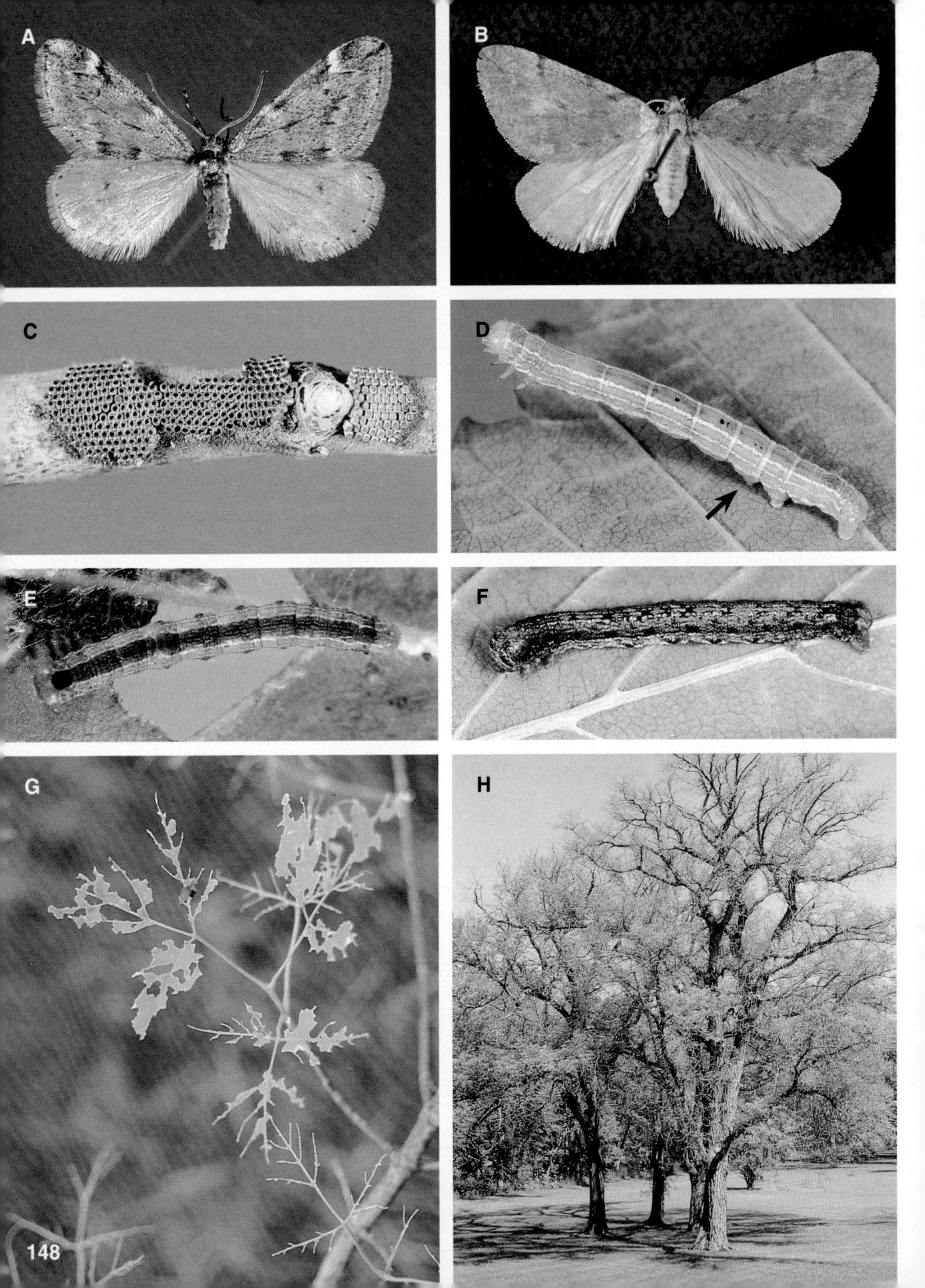
A
B
C
D
E
F
G
H

Symptoms and Signs

Larvae of fall cankerworm, ***Alsophila pometaria*** (Harris) (Lepidoptera: Geometridae), and the spring cankerworm, ***Paleacrita vernata*** (Peck) (Geometridae), defoliate trees and shrubs in urban areas and shelterbelts. Starting in late May, young larvae of both species chew small holes in developing leaves. As feeding continues, these holes gradually enlarge until all of the leaf tissue, except the larger veins and midribs, is destroyed (G). These insects do not web leaves together. Other insects cause similar damage. Thus, identification is easiest if larvae are observed. Larvae of both species crawl in a looping motion. Larvae of ***A. pometaria*** vary from light green with white lines (D) to brownish green with a dark dorsal band (E), and are about 25 mm long when fully grown. They have three pairs of prolegs, one of which is rudimentary (D). ***Paleacrita vernata*** larvae (F) have two pairs of prolegs, are mottled yellowish green to blackish, may have yellow lateral stripes, have a pair of dorsal tubercles on the posterior end of the abdomen, and are 20–30 mm long when fully grown.

Distribution and Hosts

Both species are distributed from the east coast of Canada to Alberta, and throughout most of the United States. Manitoba maple (*Acer negundo* L.) and American elm (*Ulmus americana* L.) are preferred by ***A. pometaria***, and Siberian elm (*Ulmus pumila* L.) is preferred by ***P. vernata***. Both species also attack many other hardwood species.

Life Cycle

Alsophila pometaria has one generation per year. Overwintered eggs hatch when buds swell. Larvae start feeding when foliage appears and continue until the end of June. Mature larvae drop to the ground, spin silken cocoons in the soil, and pupate. The grayish brown adults emerge in October. Females are wingless and about 12 mm long. Males have a wingspan of about 30 mm and the fore wings are semitransparent and have two irregular light bands (A). Female moths climb trees, mate, and lay eggs. Eggs are brownish, flower-pot-like in shape, and are laid in compact masses on the bark high on the trunk and branches (C).

Paleacrita vernata has one generation per year and overwinters as mature larvae in earthen cells in the duff. Pupation occurs in these cells in early spring, and adults emerge in April and early May. Female moths are wingless, vary from brownish to blackish, have two transverse rows of flattened reddish spines on each segment, and are 8–12 mm long. Males have a wingspan of about 35 mm, and their fore wings are semitransparent brownish gray with several dark brown spots and broken lines (B). Females climb trees to mate. Oval, pearl-colored eggs are laid in loose clusters in bark crevices or under bark scales on the lower parts of the trunk and lower branches. Larval feeding habits and seasonal development are similar to that of the fall cankerworm. Mature larvae drop to the ground to pupate.

Damage

Although these two species co-occur in southern parts of the prairie provinces, ***A. pometaria*** is usually the dominant species and probably causes the most damage. Severely defoliated trees (H) will generally refoliate by mid-July. Three or more consecutive years of severe defoliation may cause many of the upper branches to die, destroying the tree's aesthetic value, and may also contribute to tree mortality. When cankerworm populations are large, starving larvae in search of food may drop on silken threads and be a public nuisance around homes and recreation areas.

Figure 66. Cankerworms. A. Adult *A. pometaria* male. **B.** Adult *P. vernata* male. **C.** *A. pometaria* egg mass. Eggs in cluster to the right are unhatched. **D.** Lateral view of green form of *A. pometaria* larva. Note rudimentary proleg at arrow. **E.** Dorsal view of dark color form of *A. pometaria* larva. **F.** Lateral view of dark form of *P. vernata* larva. Note pair of tubercles at arrow. **G.** *A. pometaria* feeding damage on Manitoba maple. **H.** Elms defoliated by *A. pometaria*.

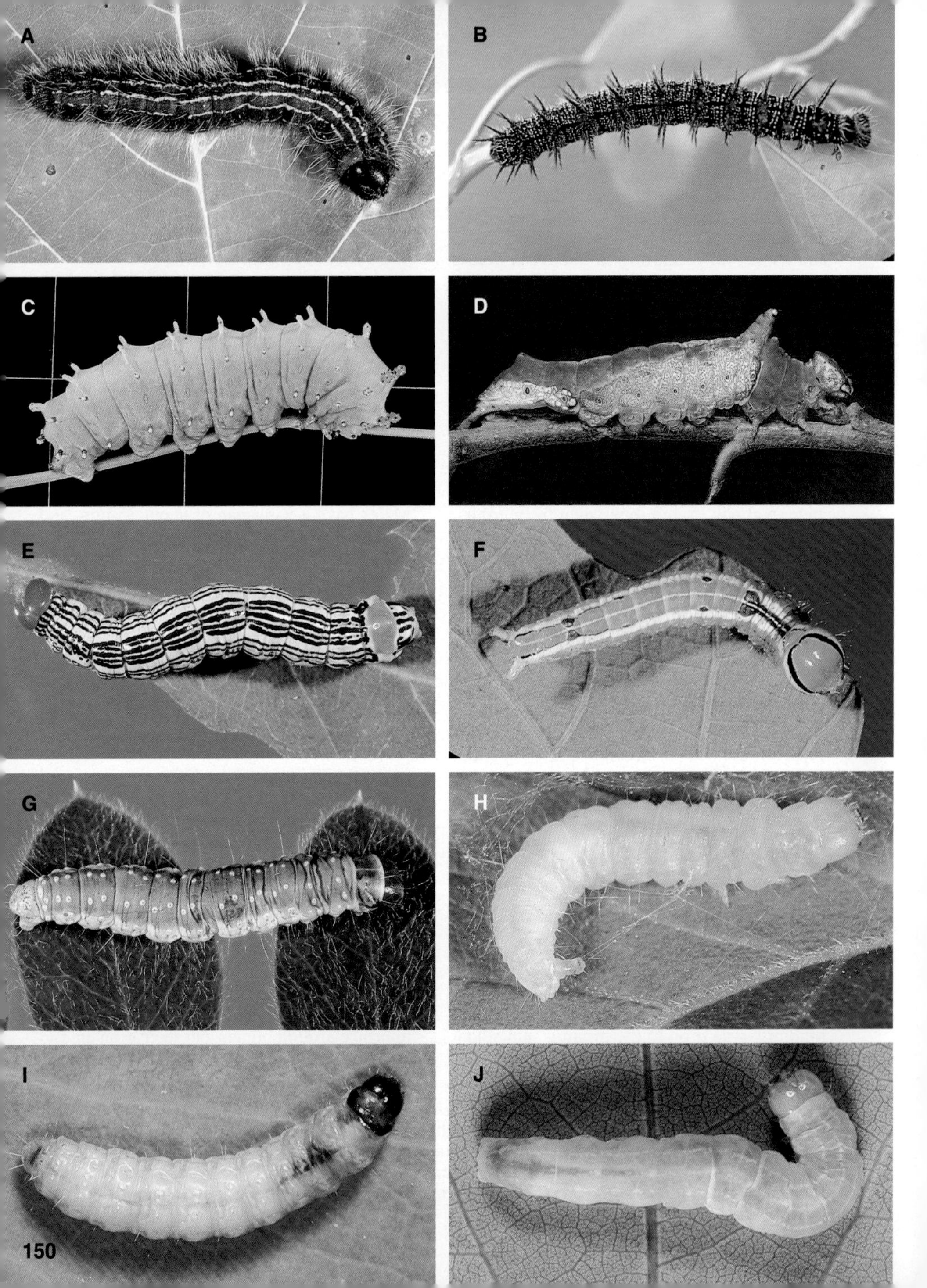
A
B
C
D
E
F
G
H
I
J

Symptoms and Signs

Of the hundreds of minor species of caterpillars (Lepidoptera) that feed on hardwoods in the prairie provinces, several cause significant localized defoliation. Damage symptoms caused by these species are not particularly diagnostic; larvae are usually required for identification. All species covered in this chapter have three pairs of thoracic legs and five pairs of abdominal prolegs.

The following six species feed openly on foliage and do not web leaves together. The larvae of most species are 35–50 mm long when fully grown. Yellownecked caterpillars, ***Datana ministra*** (Drury) (Notodontidae), have black bodies covered with white hairs and with eight longitudinal yellow lines, yellowish orange thoraxes, and black heads (A). Larvae of all ages have the habit of holding both ends of the body in an elevated position when disturbed. Mourningcloak butterfly, ***Nymphalis antiopa*** (L.) (Nymphalidae), larvae are black and covered with numerous small white dots and branched spines, and have a large conspicuous red dot on each of the first seven abdominal segments (B). Cecropia moth, ***Hyalophora cecropia*** (L.) (Saturniidae), larvae are pea-green and about 90 mm long when fully grown (C). Each segment has two prominent dorsal, bristled, reddish or yellow tubercles. The dorsal parts of the thorax and abdomen of the lacecapped caterpillar, ***Oligocentria lignicolor*** (Walker) (Notodontidae), are green and the rest of the body is mottled creamy-white and light brown (D). There is a large dorsal hump on the first abdominal segment and a smaller hump on the third thoracic segment. Redhumped oakworms, ***Symmerista canicosta*** Franclemont (Notodontidae) (E), have orange-red heads, a rounded orange hump on the eighth abdominal segment, and several black, white, and yellow longitudinal lines. The variable oak leaf caterpillar, ***Lochmaeus manteo*** Doubleday (Notodontidae), is yellowish green with white and yellow stripes and variable purplish brown bands or spots (F). The best diagnostic feature of this species is a purplish brown stripe on each side of the head.

The following four species feed while concealed within leaves that are rolled or tied together with silken webbing. Mature larvae of the first three species are 20–25 mm long. Fruit tree leafroller, ***Archips argyrospila*** (Walker) (Tortricidae), larvae are light green underneath and dark green or brownish gray with two rows of white spots on top (G). They have reddish brown to dark brown heads and thoracic shields. Larger boxelder leafroller, ***A. negundana*** (Dyar), larvae are pale green all over (H). Aspen leafroller, ***Pseudexentera oregonana*** (Walsingham) (Tortricidae), larvae have light to dark brown heads, creamy-white bodies, and brown thoracic shields (I). Aspen twoleaf tier, ***Enargia decolor*** (Walker) (Noctuidae), larvae have yellowish heads, yellowish green bodies with several longitudinal white lines, and are 25–35 mm long at maturity (J).

Distribution and Hosts

For details of distribution and hosts of these species, refer to Table 5.

Life Cycle

Datana ministra has a 1-year life cycle and overwinters as pupae in the soil. Adults emerge in June or July. Eggs are white and laid in clusters on the underside of leaves. Young larvae skeletonize leaves but older larvae eat all but the largest veins. Larval development is completed by late summer or early fall, when mature larvae fall to the ground to pupate.

Nymphalis antiopa overwinters as adults in sheltered places. Adults emerge from hibernation early in the spring and lay clusters of eggs on small twigs. On hatching, young larvae feed gregariously, initially in silken shelters and later more openly, stripping all the leaves on one branch before moving on together to another. Mature larvae suspend themselves from twigs or other objects and pupate to form a characteristic chrysalis. After emerging, adults fly for a while before seeking hibernation sites. Only one generation per year is produced in the prairie provinces.

Figure 67. Other common defoliating caterpillars. A. *Datana ministra*. **B.** *Nymphalis antiopa*. **C.** *Hyalophora cecropia*. **D.** *Oligocentria lignicolor*. **E.** *Symmerista canicosta*. **F.** *Lochmaeus manteo*. **G.** *Archips argyrospila*. **H.** *Archips negundana*. **I.** *Pseudexentera oregonana*. **J.** *Enargia decolor*.

Hyalophora cecropia has a 1-year life cycle and overwinters as pupae in cocoons attached to twigs or branches. Adults emerge in late spring to early summer. Eggs are laid in groups of 3–30 on the undersurface of leaves. Larvae feed from June to September. Mature larvae spin gray to brown spindle-shaped cocoons with pointed ends, in which they pupate.

Lochmaeus manteo, ***S. canicosta***, and ***O. lignicolor*** each have one generation per year and overwinter as pupae or prepupae on the ground. Adults begin emerging and laying eggs in June. Larvae are present from July to late September.

Archips argyrospila and ***A. negundana*** have similar life cycles. They overwinter in the egg stage on the host. Larvae hatch in spring and begin feeding on opening leaves, webbing them together with silk. Pupation occurs among webbed leaves from late June to mid-July. Adults fly throughout most of July and early August and lay eggs in clusters on the twigs of the host. There is one generation per year.

Pseudexentera oregonana pupae overwinter in the litter. Adults emerge early in the spring, often when there is still snow on the ground. Females lay flat oval eggs singly on twigs and branches. The young larvae tunnel into the expanding buds and subsequently web the expanding leaves together to form a shelter. Later they roll leaves tightly together. Mature larvae drop to the ground to pupate in August to complete a 1-year life cycle.

Enargia decolor has one generation per year and overwinters as eggs on twigs. Larvae hatch as buds break and each larva webs together two leaves and feeds inside the shelter. As larvae develop they tie other leaves together. Larvae fall to the ground to pupate in late July. Adults emerge, fly, and lay eggs in August.

Damage

Nymphalis antiopa and ***D. ministra*** occasionally cause significant localized defoliation. Widespread outbreaks of ***H. cecropia*** have occurred on shelterbelts, especially in Saskatchewan. ***Lochmaeus manteo***, ***S. canicosta***, and ***O. lignicolor*** have caused severe localized defoliation of bur oak (*Quercus macrocarpa* Michx.) and other hardwoods in southern Manitoba. Short-lived outbreaks of ***A. argyrospila*** and ***A. negundana*** have occurred in southern Manitoba and Saskatchewan. ***Enargia decolor*** and ***P. oregonana***, along with other less common species of leaf rollers and tiers, have contributed to significant defoliation of aspen throughout the prairie provinces.

Table 5. Distribution and hosts of common defoliating caterpillars of deciduous trees in the prairie provinces

Species	Distribution[a]	Hosts
Datana ministra	Transcontinental	White birch (*Betula papyrifera* Marsh.), white elm (*Ulmus americana* L.), apple (*Malus* spp.), willow (*Salix* spp.), other broadleaf trees
Nymphalis antiopa	Transcontinental	Willow, white elm, trembling aspen (*Populus tremuloides* Michx.), other broadleaf trees
Hyalophora cecropia	Nova Scotia to Alberta	Manitoba maple (*Acer negundo* L.), ash (*Fraxinus* spp.), elm (*Ulmus* spp.), other broadleaf trees
Lochmaeus manteo	Nova Scotia to Manitoba	Oak (*Quercus* spp.), birch (*Betula* spp.), other broadleaf trees
Symmerista canicosta	Nova Scotia to Manitoba	Oak
Oligocentria lignicolor	Nova Scotia to Manitoba	Birch, oak
Archips argyrospila	Transcontinental	Fruit trees, aspen, willow, other broadleaf trees
Archips negundana	Manitoba to British Columbia	Manitoba maple
Pseudexentera oregonana	Ontario to British Columbia	Trembling aspen
Enargia decolor	Transcontinental	Poplar, white birch, willow

[a] Species are also distributed in adjacent parts of the United States.

A
B
C
D
E
F
G
H
I
J

Symptoms and Signs

The sawfly species (Hymenoptera: Tenthredinidae) treated in this chapter are among the most common, conspicuous, or destructive (or potentially destructive) in the prairie provinces. Mature larvae are needed for correct identification of these species because damage symptoms are not diagnostic. All of the following sawfly species feed openly on leaves, consuming all of the leaf tissue except large veins and midribs (C), and they do not web leaves together. Young sawfly larvae are generally difficult to identify, so only mature larvae are described here. Sawfly larvae have three pairs of thoracic legs and at least six pairs of prolegs. ***Nematus calais*** Kirby larvae have black heads, grayish green bodies with numerous black spots, a black triangular middorsal anal patch, and a row of pale yellow spots along each side of the body (D). Larvae of ***N. ventralis*** Say are mainly black with a row of large bright yellow spots along each side of the body (E). Larvae of ***N. limbatus*** Cresson have black heads and bodies that are grayish green with diffuse yellow patches and black spots (F). Mature ***Nematus*** larvae are 18–22 mm long. Elm sawfly, ***Cimbex americana*** Leach, larvae are yellowish green with a middorsal bluish and black stripe, and are about 55 mm long at maturity (H). Birch sawfly, ***Arge pectoralis*** (Leach), larvae have reddish heads, black thoracic legs, yellowish green bodies with numerous black spots, and a black dorsal anal patch, and are about 25 mm long when fully grown (I). Mature mountain ash sawfly, ***Pristiphora geniculata*** (Hartig), larvae have yellow heads and thoracic legs, yellow bodies with black spots, and a pinkish dorsal anal patch, and are 12–13 mm long (J).

Distribution and Hosts

The ***Nematus*** species are distributed throughout southern Canada and the northern United States, east of the Rocky Mountains. All three species feed on willow (*Salix* spp.) and ***N. ventralis*** also feeds on poplar (*Populus* spp.). ***Arge pectoralis*** and ***C. americana*** have transcontinental distributions; ***A. pectoralis*** feeds on birch (*Betula* spp.) and ***C. americana*** feeds mainly on elm (*Ulmus* spp.) and willow, but also many other hardwoods. ***Pristiphora geniculata*** was introduced from Europe into eastern North America and has spread as far west as southeastern Manitoba. It feeds on mountain ash (*Sorbus* spp.).

Life Cycle

All species overwinter as prepupae in tough, leathery cocoons in the duff or soil. ***Cimbex americana*** and ***A. pectoralis*** have one generation per year and the other species have two per year. After winter, adults emerge and reproduce in late May and early June. For those species with a second generation, adults of the next generation emerge and reproduce in late July and early August. Adults of all species are stout-bodied, and have four membranous wings; the basic body color is black with yellowish to yellowish orange marks on the thorax or abdomen (A, G). ***Cimbex americana*** adults are 18–28 mm long and have clubbed antennae (G). Adults of ***A. pectoralis*** are 8–12 mm long, those of ***Nematus*** are 6–9 mm long, and those of ***P. geniculata*** are 5–7 mm long. All have straight antennae (A). Eggs are laid in slits cut in the leaf surface (B). Young larvae tend to be gregarious (C) but become more solitary as they get larger.

Damage

Severe outbreaks of ***P. geniculata*** have occurred on mountain ash in eastern Canada. The other species can also cause significant localized defoliation. The effects of defoliation depend on tree health before defoliation, amount of foliage removed, and the number of consecutive years of defoliation. Prolonged infestations may cause reduction in stem radial growth, branch and twig mortality, loss of aesthetic value, but rarely tree death.

Figure 68. Common defoliating sawflies. A–D. *Nematus calais*. **A.** Adult. **B.** Egg slits on a willow leaf. **C.** Young larvae on willow. **D.** Mature larva. **E.** *N. ventralis* larva. **F.** *N. limbatus* larva. **G, H.** *Cimbex americana*. **G.** Adult. **H.** Larva. **I.** *Arge pectoralis* larva. **J.** *Pristiphora geniculata* larva.

A
B

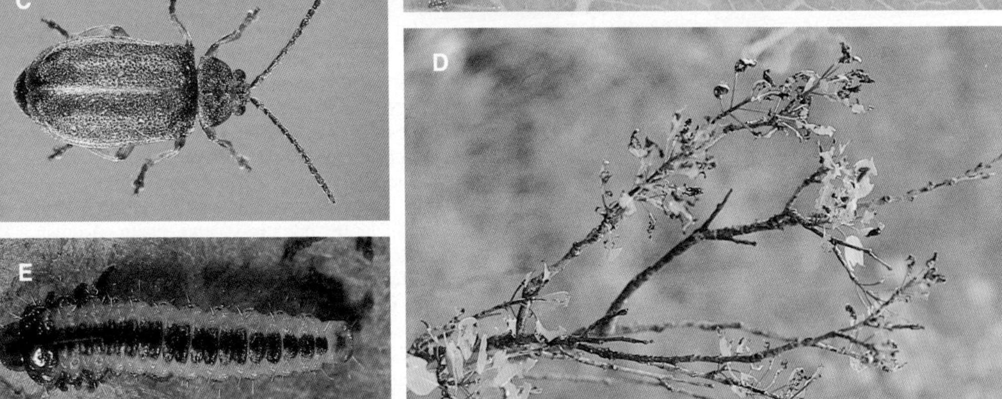
C
D
E

F
G

Symptoms and Signs

Larvae of the gray willow leaf beetle, ***Tricholochmaea decora*** (Say) (Coleoptera: Chrysomelidae), skeletonize willow leaves by eating all but the upper epidermis and veins (F), giving trees a scorched brown appearance (G). This symptom first appears in late June. Small larvae of the American aspen beetle, ***Gonioctena americana*** (Schaeffer) (Chrysomelidae), also skeletonize leaves; however, large larvae may consume entire leaves except for major veins and midribs, giving the foliage a ragged appearance (D). Larvae and adults of each of these species feed on foliage and both stages may be seen feeding together. Adults of ***T. decora*** are grayish to yellowish brown, occasionally almost black (C), and are about 5 mm long. Larvae have black heads, light yellow bodies with a broad black middorsal stripe and a broken black lateral line on each side, and are 8–10 mm long when fully grown (E). Adults of ***G. americana*** are 5–6 mm long and variable in color. The elytra of some specimens are red or reddish yellow with black spots (A); others are almost entirely dark brown or black. Larvae are black and 8–10 mm long when fully grown (B). Adults or larvae are usually present on foliage from June to mid-August.

Distribution and Hosts

Both species have a transcontinental distribution. Adults and larvae of ***G. americana*** feed on trembling aspen (*Populus tremuloides* Michx.). Adults of ***T. decora*** feed on aspen and willow (*Salix* spp.), but larval feeding is apparently confined to willow.

Life Cycle

Tricholochmaea decora has one generation per year. Sexually immature adults overwinter in the litter. Adults emerge in the spring and feed gregariously and extensively on foliage until mid-June. Oviposition occurs on willow from mid-June to early July. Eggs are laid in clusters covered with fecal matter at the base of willow bushes, usually on rough bark. Larvae feed until late July or early August, when they drop to the duff to pupate. When adults emerge about 2 weeks later, they feed briefly on foliage, then enter the litter to hibernate.

Gonioctena americana has one generation per year. Sexually immature adults overwinter in the litter. Adults emerge from hibernation sites when aspen foliage is fully developed. They feed by chewing notches in the edges of leaves for several days. After mating, females lay clusters of eggs, containing fully developed embryos, on the underside of leaves. Larvae emerge almost immediately and feed gregariously when young. Feeding lasts 3–4 weeks, and mature larvae drop to the litter or soil in late June to pupate. Adults emerge in mid-July, feed on foliage for a while, then drop to the ground to hibernate.

Damage

These species do not appear to cause serious injury to their hosts, even during severe infestations. Feeding damage on trees renders them unsightly (G) and likely causes reduction in radial growth of stems. Outbreaks usually subside before any appreciable branch or tree mortality occurs. Defoliation of ornamental and shade trees reduces their aesthetic value.

Figure 69. Gray willow leaf beetle and American aspen beetle. A. *Gonioctena americana* adult. **B.** *G. americana* larvae feeding on an aspen leaf. **C.** *Tricholochmaea decora* adult. **D.** *G. americana* feeding damage on aspen. **E.** *T. decora* larva. **F.** *T. decora* feeding damage on lower surface of willow leaf. **G.** Willows defoliated by *T. decora*.

A
B
C
D
E
F
G
H
I
J

OTHER COMMON LEAF BEETLES

Symptoms and Signs

Trees defoliated by leaf beetles (Coleoptera: Chrysomelidae) often have a scorched appearance or the foliage appears thin and ragged. Damage symptoms caused by leaf beetles can usually be distinguished from those caused by defoliating sawflies and caterpillars. Most adult leaf beetles chew many small holes through leaves (H). Larvae skeletonize leaves by consuming all leaf tissue except the midrib and most veins to give the leaves a lacelike appearance (E, I, J). Caterpillars and sawfly larvae that skeletonize leaves usually leave only the midribs and largest veins, or they leave a layer of epidermis covering the veins. It is difficult to discriminate among leaf beetle species using damage characteristics; observation of larvae and adults is desirable.

Chrysomela adults are oval and 6–10 mm long. Adult cottonwood leaf beetles, ***C. scripta*** Fabricius, have black heads, pronota that are black in the center and reddish on the sides, and yellowish elytra with black spots and lines (A). Adult aspen leaf beetles, ***C. crotchi*** Brown, have black heads and pronota and the elytra are brownish yellow with no markings (D). Adults of a related species, ***C. falsa*** Brown, are similar to those of *C. scripta* but have brownish yellow elytra with a different pattern of black spots (F). Adults of ***Phratora*** and ***Altica*** are oval, 5–6 mm long, and shiny steel-blue, coppery, purplish, or black in color (G, H). ***Altica*** adults jump readily when disturbed, and are thus called flea beetles. Adult ***Phratora*** can be separated from flea beetles by the rows of punctures in their elytra (G). Leaf beetle larvae have thoracic legs but no prolegs. Larvae of ***Chrysomela*** species are similar and have black heads, brownish thoracic shields, yellowish bodies with two rows of black spots along the back, and dark eversible glands along each side of the abdomen (C, D). Larvae of ***Phratora*** and ***Altica*** are black (I) and difficult to identify to species.

Distribution and Hosts

Chrysomela falsa and ***C. crotchi*** have transcontinental distributions and ***C. scripta*** occurs from British Columbia to Ontario and in adjacent parts of the United States. Most ***Phratora*** and ***Altica*** species on hardwood trees in the prairie provinces are widely distributed. ***Chrysomela scripta*** attacks various poplars (*Populus* spp.), particularly hybrids, and willows (*Salix* spp.), but not aspen (*Populus tremuloides* Michx.). ***Chrysomela crotchi*** feeds mainly on aspen, but sometimes other poplars, and ***C. falsa*** attacks balsam poplar (*P. balsamifera* L.). ***Phratora*** and ***Altica*** species are most commonly found on poplars, willows, birches (*Betula* spp.), and alder (*Alnus* spp.).

Life Cycle

Chrysomela species overwinter as sexually immature adults in the duff. Adults emerge when leaves have flushed, feed for a while, and reproduce. Eggs are variously colored and laid in clusters on leaves (B). Larvae develop through three or four instars. Pupation occurs on the foliage. ***Chrysomela scripta*** has two generations per year; ***C. crotchi*** has only one. The life cycles of many species of ***Phratora*** and ***Altica*** are not known, but some species have a life cycle similar to that described above.

Damage

Chrysomela scripta prefers young foliage. High populations often kill shoots and cause deformed growth. Damage is especially prevalent in plantations of hybrid poplars and in nursery stooling beds, where yield of cuttings is seriously reduced. ***Chrysomela crotchi*** occasionally causes conspicuous but localized and short-lived defoliation of aspen, but likely has little impact. ***Phratora*** and ***Altica*** feeding causes trees and shrubs to look unsightly but causes little permanent damage.

Figure 70. Other common leaf beetles. A–C. *Chrysomela scripta*. **A.** Adult. **B.** Eggs. **C.** Larvae. **D, E.** *C. crotchi*. **D.** Adult (left) and larva (right). **E.** Feeding damage on aspen. **F.** *C. falsa* adult. **G.** *Phratora* adult. **H.** *Altica* adult. **I, J.** *Altica ambiens*. **I.** Larvae and feeding damage on alder. **J.** Skeletonized alder leaves.

A
B
C
D
E
F
G

Symptoms and Signs

Larvae of the pear sawfly, ***Caliroa cerasi*** (L.) (Hymenoptera: Tenthredinidae), and the birch skeletonizer, ***Bucculatrix canadensisella*** Chambers (Lepidoptera: Lyonetiidae), skeletonize leaves by eating all tissue except for the midrib, veins, and a layer of epidermis (B). The first indications of the presence of these species are yellowish spots on host foliage. Young ***C. cerasi*** larvae produce these spots by feeding on the upper surface of leaves. Young ***B. canadensisella*** larvae initially cause these spots by mining the leaves; they eventually emerge to feed on the undersurface of leaves. As feeding by both species progresses, affected areas on individual leaves enlarge and merge, until the leaves look bleached (B). The foliage eventually turns brown to reddish brown in late July and August, to give the trees a scorched appearance (G). Damage symptoms may be similar to those caused by some skeletonizing leaf beetles. Therefore observation of damaging agents facilitates easier identification. Leaves damaged by ***B. canadensisella*** have small, white silken pads (molting tents) attached to them (F), which differentiate them from leaves similarly damaged by other species. Larvae of ***C. cerasi*** are shiny and sluglike, they change from a blackish (B) to a yellowish green (C) as they mature, and they are about 11 mm long when fully grown. Young ***B. canadensisella*** larvae are white and legless when they mine the leaf tissue. Older larvae have functioning thoracic legs, five pairs of abdominal prolegs, brown heads, and yellowish green bodies that are distinctly segmented, and they are about 6 mm long when fully grown (E).

Distribution and Hosts

Caliroa cerasi was introduced into North America from Europe. In Canada its preferred hosts are *Cotoneaster* spp., hawthorn (*Crataegus* spp.), mountain ash (*Sorbus* spp.), pin cherry (*Prunus pensylvanica* L.f.), and various other fruit-bearing trees. ***Bucculatrix canadensisella*** has a transcontinental distribution. It prefers white birch (*Betula papyrifera* Marsh.) but also attacks other birch species.

Life Cycle

Caliroa cerasi larvae overwinter in small cocoons in the soil and pupate in the spring. The approximately 5-mm-long, black sawfly adults (A) emerge from mid-June until mid-July. Eggs are deposited in slits cut by the female in the lower surface of leaves. Larvae hatch in about 2 weeks and move to the upper surface of leaves to feed. Mature larvae drop to the ground and spin silken cocoons in the litter or soil, where some pupate immediately and emerge as adults. If the weather is suitable, this second generation may complete its development before fall.

Bucculatrix canadensisella has one generation per year and overwinters as pupae in silken cocoons spun in the litter or soil. Adult moths emerge from mid-June to mid-July and reproduce. They have a wingspan of about 7 mm and are brown with white bands on the fore wings (D). Eggs are laid singly on either side of the leaves and hatch in about 2 weeks. Young larvae mine leaves and form narrow winding tunnels. They emerge from the tunnels 3–4 weeks later to feed on the undersurface of leaves. Larvae subsequently molt twice in tiny white silken tents on the leaf surface (F). Mature larvae drop to the litter and spin light brown, ribbed, silken cocoons (F), in which they pupate and overwinter.

Damage

The scorched appearance of trees reduces their aesthetic value. Healthy trees can withstand several years of moderate to severe attack because damage occurs in late summer, near the end of the growing season. Severe attack by ***B. canadensisella*** in consecutive years may kill some twigs in the upper crown and slightly reduce stem radial growth.

Figure 71. Leaf skeletonizers. A–C. *Caliroa cerasi*. **A.** Adult. **B.** Young larvae and damage on hawthorn. **C.** Mature larva. **D–G.** *Bucculatrix canadensisella*. **D.** Adult. **E.** Larva. **F.** Molting tent (bottom) and cocoon (top). **G.** Damaged white birch.

A
B
C
D
E
F
G
H

Symptoms and Signs

Damage by the birch leafminer, ***Fenusa pusilla*** (Lepeletier) (Hymenoptera: Tenthredinidae), first becomes noticeable in early June, when small, light green or gray spots appear on the leaves where eggs were deposited (D). Oviposition by the ambermarked birch leafminer, ***Profenusa thomsoni*** (Konow) (Tenthredinidae), causes similar symptoms in July (H). When eggs hatch and larvae begin feeding, these spots develop into brown blotches (F), usually several per leaf. As larvae feed between the leaf surfaces, the blotches increase in size and merge, eventually covering most of the leaf (E). Larvae feeding in the blotches are whitish, slightly flattened in appearance, and 6–7 mm long when fully grown (A, B). Variations in the black marks on the underside of larvae can be used to discriminate among species. ***Profenusa thomsoni*** is the most common species of birch leafminer in the prairie provinces. A third species of birch leaf-mining sawfly, the late birch leaf edgeminer, ***Heterarthrus nemoratus*** (Fallen) (Tenthredinidae), is rare in the prairie provinces.

Distribution and Hosts

All three species were accidentally introduced from Europe to North America early in the 20th century and are now widely distributed in Canada and the northern United States. All native and exotic birches (*Betula* spp.) are susceptible to damage by at least one species of birch leafminer.

Life Cycle

Fenusa pusilla is the first species to attack birch in the spring. Adults are small, stout, black insects with two pairs of membranous wings, and are 3–4 mm long (C). In mid- to late May, females each lay up to 20 eggs in slits near the midribs on the upper surfaces of young leaves in the sun-exposed parts of the crown. Larvae hatch from the eggs in early June and feed on the tissue between the leaf surfaces. Mature fifth-instar larvae do not feed, but emerge from the leaves in late June to mid-July and drop to the ground, where they construct earthen cells (cocoons) beneath the soil surface. Most larvae likely remain in the earthen cells to overwinter, with pupation occurring the next spring. However, a small second generation sometimes occurs, which indicates that some first-generation larvae may pupate immediately and emerge as adults about a month later to oviposit. Larvae of this second generation fall to the ground in late August.

Profenusa thomsoni has one generation per year. Adults resemble those of *F. pusilla*. The parthenogenetic females emerge in July and lay eggs along the veins in the basal and central area of the upper surface of the leaves. No males of this species have been found in North America. There are five feeding larval instars and a sixth nonfeeding instar. Sixth-instar larvae emerge from leaves in late August and drop to the soil to construct small earthen cells, in which they overwinter. Pupation occurs in early summer of the next year.

Damage

Although most of the inner leaf tissue may be destroyed, tree health is not usually seriously affected, since a vigorous tree can withstand many years of light to moderate damage. Extensive mining of leaves can stress the birch, however, and combined with lack of adequate moisture and attack by other insects or diseases, may cause branch and top dieback. This dieback, combined with leaf browning (G), reduces aesthetic value of birches in ornamental settings.

Figure 72. Birch leafminers. A. *Fenusa pusilla* larva. **B.** *Profenusa thomsoni* larva. **C.** *F. pusilla* adult. **D.** *F. pusilla* egg punctures in leaf of white birch (*Betula papyrifera* Marsh). **E.** White birch leaves mined by *F. pusilla* and *P. thomsoni* larvae. **F.** *F. pusilla* mine in white birch leaf. **G.** White birch trees severely attacked by *P. thomsoni*. **H.** *P. thomsoni* egg punctures and early larval mining on birch leaf.

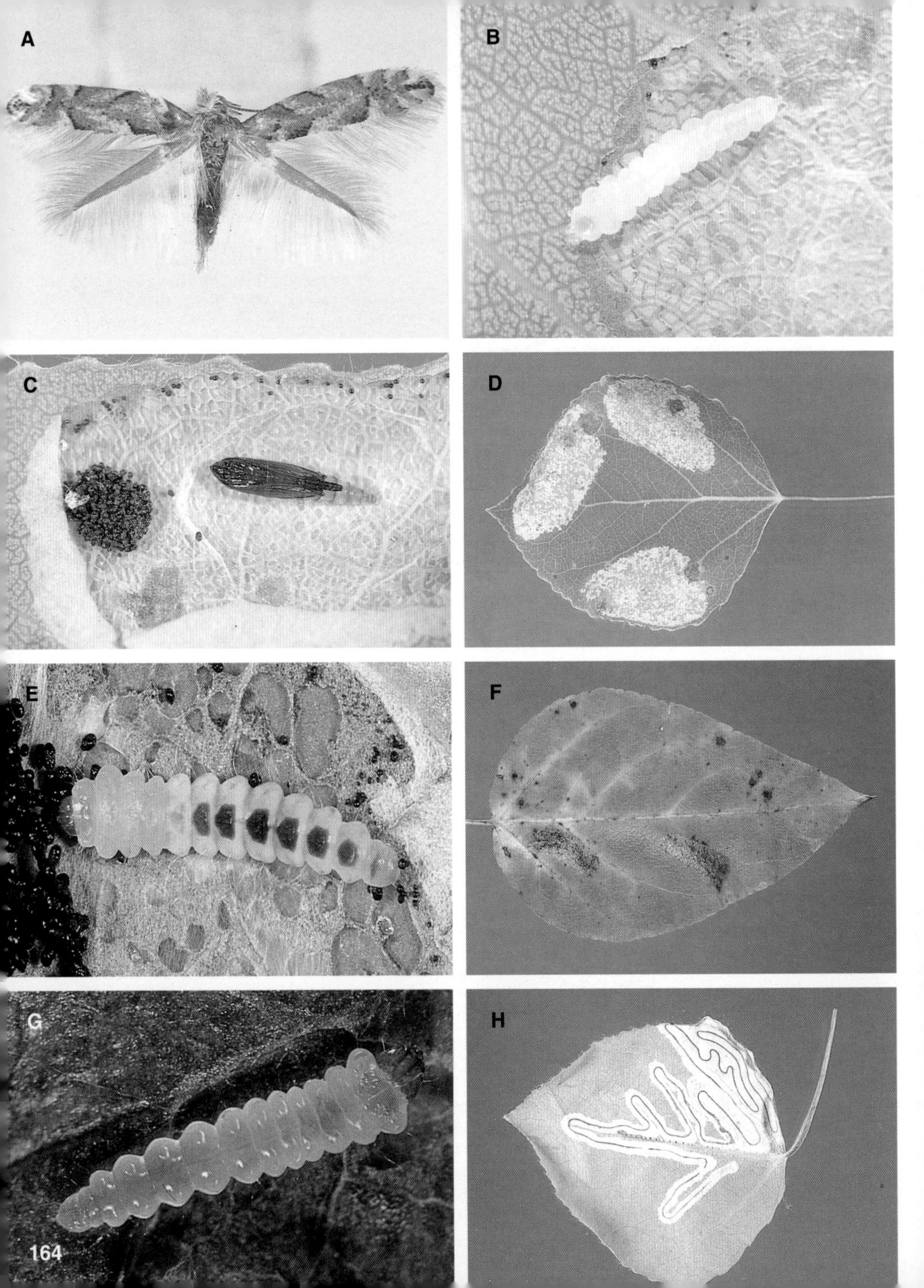
A
B
C
D
E
F
G
H

Symptoms and Signs

Among the most conspicuous leafminers in the prairie provinces are ***Phyllonorycter*** nr. ***salicifoliella*** (Chambers) (Lepidoptera: Gracillariidae); ***P.*** nr. ***nipigon*** (Freeman); the willow leafminer, ***Micrurapteryx salicifoliella*** (Chambers) (Gracillariidae); and the aspen serpentine leafminer, ***Phyllocnistis populiella*** Chambers (Gracillariidae). The larvae of the first three of these moth species construct blotch mines (D, F) and ***P. populiella*** constructs serpentine mines (H). The mines of all four species are conspicuous from about mid-June onwards. The foliage of trees that have been heavily attacked appears bleached yellowish to brownish. Other species of insects cause blotch mines in poplars; however, the two species of ***Phyllonorycter*** covered in this chapter may be identified by the color of the blotches, the color and shape of larvae, and the presence of pupae in the mines. Mature larvae of ***P.*** nr. ***salicifoliella*** are about 6 mm long with distinctly segmented bodies; the front half of the body (including the head) is creamy-white, and the back half is orange-yellow (B). Larvae of ***P.*** nr. ***nipigon*** are similar to those of *P.* nr. *salicifoliella* except that each abdominal segment has a black dorsal spot and the head is brown (E). Pupae of ***P.*** nr. ***salicifoliella*** are orange-yellow with blackish heads and wing pads (C), but those of ***P.*** nr. ***nipigon*** are totally black. Mature larvae of ***M. salicifoliella*** are about 7 mm long; they have brown heads and distinctly segmented bodies that are creamy-white, except for the thorax and first one or two abdominal segments, which are pale yellow (G). Pupae are similar in color to those of *P.* nr. *salicifoliella*.

Distribution and Hosts

Phyllocnistis populiella has a transcontinental distribution and attacks trembling aspen (*Populus tremuloides* Michx.). Information on the distribution of the other species is scanty because the taxonomy of these species is poorly understood. All appear to be widely distributed in the prairie provinces. ***Phyllonorycter*** nr. ***nipigon*** attacks balsam poplar (*Populus balsamifera* L.); ***P.*** nr. ***salicifoliella*** attacks aspen; and ***M. salicifoliella*** attacks willows (*Salix* spp.).

Life Cycle

The life cycles of all species are not well known. All apparently have one generation per year and overwinter as adults under bark scales or in the litter. Adults of ***Phyllocnistis populiella*** are tiny silvery-white moths with a wingspan of about 5 mm. The other three species are light to dark brown, have fore wings with silvery-white spots or stripes, and have a wingspan of 8–9 mm (A). All species have a fringe of long hairs around the wings. Adults emerge from overwintering sites when the leaves are fully formed and lay eggs on the leaf surfaces. Larvae bore into the leaves and feed on parenchyma tissue. Fourth-instar larvae of ***Phyllonorycter*** spin a heavy layer of silk on the inside of the leaf epidermis. The silk contracts as it dries, creating a bulge typical of blotch miners. All four species pupate within the mines in July and adults emerge from mid-July to mid-August. Adults are active for a short time before hibernating.

Damage

Conspicuous infestations of all four species have been reported in the prairie provinces on occasion. However, these outbreaks tend to be short-lived and have little impact on trees. Foliage discoloration and associated premature leaf drop may reduce aesthetic value of trees in parkland settings.

Figure 73. Poplar and willow leafminers. A–D. *Phyllonorycter* nr. *salicifoliella.* **A.** Adult. **B.** Larva in mine on aspen leaf. **C.** Pupa and frass in mine on aspen leaf. **D.** Mines in aspen leaf. **E, F.** *Phyllonorycter* nr. *nipigon.* **E.** Larva in mine on balsam poplar leaf. **F.** Mines in balsam poplar leaf. **G.** *Micrurapteryx salicifoliella* larva. **H.** Aspen leaf with serpentine mine constructed by larva of *Phyllocnistis populiella.*

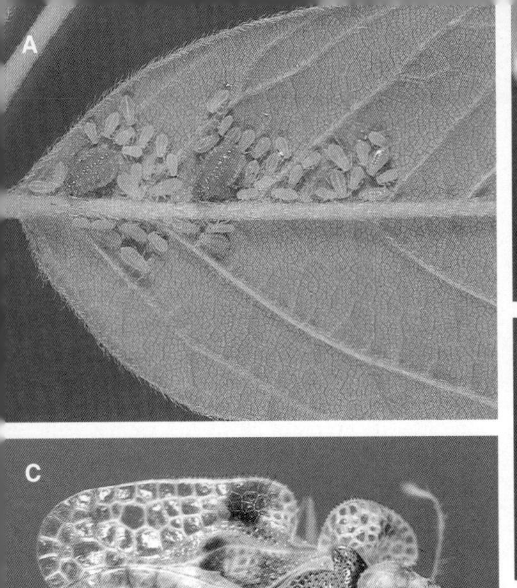
A
C

B

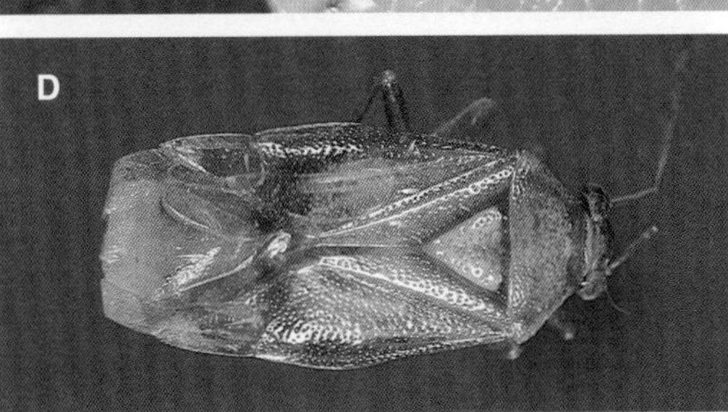
D

E

F

G

Symptoms and Signs

The species of aphids, lace bugs, and plant bugs covered in this chapter are among the most common on hardwood foliage in the prairie provinces. When these insects insert their mouthparts into leaves to suck out the fluids, the leaf tissue around the feeding sites dies and turns yellowish, giving leaves a mottled appearance (F, G). The first obvious symptom of heavy aphid infestation is often the presence of sticky honeydew coating everything under the trees. Aphids are legged, soft-bodied, 2- to 4-mm-long insects that usually occur in large aggregations (A, B). Some aphids have two pairs of membranous wings during part of their life cycle, but most are wingless. Nymphs are miniature versions of the adults (A). Boxelder aphids, ***Periphyllus negundinis*** (Thomas) (Homoptera: Aphididae), are greenish yellow (A), and smoky-winged poplar aphids, ***Chaitophorus populicola*** Thomas (Aphididae), are blackish (B).

Plant bugs and their relatives have two pairs of wings that are usually folded against the upper body. The top pair of wings (hemielytra) are mostly leathery and colored, except for the membranous tips (D, E). They cover a pair of membranous wings. Lace bug adults, such as those of the oak lace bug, ***Corythucha arcuata*** (Say) (Hemiptera: Tingidae), are about 3.5 mm long and the hemielytra have a lacelike pattern (C). They usually occur in dense aggregations. Adult ash plant bugs, ***Tropidosteptes amoenus*** Reuter (Hemiptera: Miridae), are 5–6 mm long and tan in color with pinkish markings (D). Adult boxelder bugs, ***Leptocoris trivittatus*** (Say) (Hemiptera: Rhopalidae), are black with red markings and about 12 mm long (E).

Distribution and Hosts

Periphyllus negundinis and ***L. trivittatus*** occur from Alberta to southern Ontario and in the adjacent United States, and feed on Manitoba maple (*Acer negundo* L.). ***Chaitophorus populicola*** is nearly transcontinental in distribution and feeds on poplars (*Populus* spp.) and willows (*Salix* spp.). The genus ***Corythucha*** has a transcontinental distribution and wide host range; individual species usually have more limited geographical and host ranges. ***Corythucha arcuata*** feeds on bur oak (*Quercus macrocarpa* Michx.). ***Tropidosteptes amoenus*** occurs from eastern Canada to Alberta and in the adjacent United States; it feeds on ash (*Fraxinus* spp.).

Life Cycle

Periphyllus negundinis and Chaitophorus populicola have similar life cycles. Eggs overwinter in bark crevices, then hatch in the spring. Nymphs develop into parthenogenetic, wingless females. This is followed by several successive generations of winged and wingless parthenogenetic females throughout the summer. Winged forms may disperse to new hosts or new places on the same host. Male and female aphids are produced in late summer. After mating, females lay eggs, which overwinter.

Corythucha arcuata and other lace bugs have two generations per year in the prairie provinces and overwinter as adults under loose bark or leaves. Eggs are laid in clusters on the underside of leaves in spring. Nymphs feed gregariously and develop through five instars. First-generation adults appear in midsummer and the second generation in late summer.

Tropidosteptes amoenus overwinters as eggs in bark crevices and has two generations per year. ***Leptocoris trivittatus*** overwinters as adults in sheltered places.

Damage

Damage caused by these insects is difficult to assess, but is likely not very significant. Severe infestations cause mottling of foliage and premature leaf drop, which may reduce the aesthetic value of ornamental and shade trees. Honeydew produced by large aphid infestations coats everything beneath the trees and is a public nuisance.

Figure 74. Aphids, lace bugs, and plant bugs. A. *Periphyllus negundinis* adults and nymphs on Manitoba maple. **B.** *Chaitophorus populicola* adults and nymphs on poplar. **C.** *Corythucha arcuata* adult. **D.** *Tropidosteptes amoenus* adult. **E.** *Leptocoris trivittatus* adult. **F.** *C. arcuata* damage to oak leaves. **G.** *T. amoenus* damage to ash leaves.

A
B
C
D
E

Symptoms and Signs

Many insects and mites cause galls on hardwood foliage and buds. The shape, color, and location of galls, and the host tree species can help identify the cause. Feeding of larvae of the boxelder leaf gall midge, ***Contarinia negundifolia*** Felt (Diptera: Cecidomyiidae), and the ash midrib gall midge, ***C. canadensis*** Felt, causes fleshy galls, 10–30 mm in diameter, to form along the midrib and large veins of leaves. The most visible symptom of infestation by these insects, however, is the subsequent deformation of leaves (A, B). Infested ash leaflets often turn blackish. For both species, the fleshy galls may be seen by opening the curled leaves. Larvae may be observed by slicing open the galls. Larvae are white to pale yellow, with no distinct head capsules or legs, and are about 2 mm long (C). The woolly elm aphid, ***Eriosoma americanum*** (Riley) (Homoptera: Aphididae), causes curling of elm leaves (D). Aphids produce copious honeydew, which coats everything under infested trees. Aphids are legged, soft-bodied, and dark grayish black (E). Adults are 3–4 mm long when fully grown and secrete a powdery white wax. The woolly apple aphid, ***E. lanigerum*** (Hausmann), causes rosette-shaped deformities, but is less common than *E. americanum* in the prairie provinces.

Distribution and Hosts

Contarinia canadensis and ***C. negundifolia*** are widely distributed in the northern United States and southern Canada, east of the Rocky Mountains, and feed only on ash (*Fraxinus* spp.) and Manitoba maple (*Acer negundo* L.), respectively. Both species of ***Eriosoma*** appear to have nearly transcontinental distributions. ***Eriosoma americanum*** alternates between elm (*Ulmus* spp.) and saskatoon (*Amelanchier alnifolia* Nutt.), and ***E. lanigerum*** alternates between elm and apple (*Malus* spp.), mountain ash (*Sorbus* spp.), or hawthorn (*Crataegus* spp.).

Life Cycle

Contarinia negundifolia has one generation per year and overwinters as pupae in the soil. Adults emerge when Manitoba maple leaves flush. Adults are 2-mm-long, brownish, two-winged flies. Mated females insert their eggs between the edges of young, folded leaflets that are 6–20 mm in length. Larval feeding stimulates gall formation on the midribs or lateral veins. Mature third-instar larvae drop to the ground, spin cocoons in earthen cells, and pupate in the fall. The life cycle of ***C. canadensis*** is probably similar to that of *C. negundifolia*. Eggs are laid in the veins of unfurling ash leaflets. Larval feeding stimulates formation of fleshy galls along the midribs and lateral veins.

Eriosoma americanum has several generations per year and overwinters as eggs deposited in bark crevices on twigs. Eggs hatch in the spring to yield only female nymphs, which feed on foliage and cause leaves to curl. Nymphs develop into parthenogenetic adult females, which produce a powdery white wax and honeydew. Winged female adults are produced in early summer and fly to saskatoon, where they feed and reproduce for several generations, and secrete copious amounts of white flocculence. Aphids can maintain themselves indefinitely on saskatoon, and have lost the ability to infest elm in parts of its range. A winged generation of females develops in the fall, and these fly back to elm, where they give birth to males and females. The tiny, mouthless females mate, and each deposits a single egg, which overwinters. The life cycle of ***E. lanigerum*** is similar to that of *E. americanum*.

Damage

None of these insects seriously damage their hosts. Heavy infestations reduce the aesthetic value of ornamental trees, and the honeydew produced by aphids is often a public nuisance.

Figure 75. Galled and deformed leaves. A. Ash leaflets infested by *Contarinia canadensis*. **B, C.** *C. negundifolia*. **B.** Infested Manitoba maple leaves. **C.** Larvae. **D, E.** *Eriosoma americanum*. **D.** Infested elm leaves. **E.** Female and nymphs.

A
B
C
D
E
F
G

Symptoms and Signs

Many insects and mites cause galls on hardwood foliage and buds. The shape, color, and location of galls, and the host tree species can help identify the cause. Many species of mites in the family Eriophyidae cause galls on the leaves, buds, and flowers of hardwoods. The poplar bud gall mite, ***Aceria parapopuli*** Keifer (Acari: Eriophyidae), causes cauliflowerlike galls to form by bud proliferation (C, D). These dark green, succulent galls harden and turn brick-red by late summer. They are usually confined to the lower branches. Galls remain active for 1–4 years and may grow to 4 cm in diameter. Old or abandoned galls are grayish, have a hard, ridged and furrowed surface, and may persist on trees for 5 or more years. ***Aceria*** nr. ***dispar*** (Nalepa) distorts aspen leaves (B). These galls are initially greenish but turn reddish brown as they age. The ash flower gall mite, ***A. fraxiniflora*** (Felt), causes galls in the male flowers of ash (E). Galls formed by ***A.*** nr. ***dispar*** and ***A. fraxiniflora*** are active for one summer and usually do not persist for longer than a year after they are vacated by mites. ***Aceria ulmi*** (Garman) causes many small, pale green, elongate, fingerlike galls on the blades of elm leaves (F). Some ***Eriophyes*** (Eriophyidae) mite species cause similar galls on other hardwoods; others cause small, pale green, saclike leaf galls (G). The mites contained within all of these galls are not visible to the naked eye. They are whitish to reddish, cigar-shaped, and 0.1–0.2 mm long (A).

Distribution and Hosts

The ***Aceria*** species appear to be widely distributed in the northern United States and southern Canada, east of the Rocky Mountains. The genus ***Eriophyes*** is transcontinental, but individual species may have more limited distributions. ***Aceria*** nr. ***dispar*** attacks aspen (*Populus tremuloides* Michx.); ***A. parapopuli*** attacks poplars (*Populus* spp.), especially hybrids; ***A. fraxiniflora*** attacks ash (*Fraxinus* spp.); and ***A. ulmi*** attacks white elm (*Ulmus americana* L.). The foliage of most hardwood tree species is attacked by at least one species of ***Eriophyes***.

Life Cycle

Aceria parapopuli eggs, nymphs, and larvae overwinter in galls, and begin developing in the spring. A generation is completed every 10–21 days and there are as many as eight generations per year. Mites disperse from existing galls when buds are expanding. Some crawl and others are blown to new feeding sites on buds and expanding leaves, where new galls are formed. Little is known about the life cycles of ***A.*** nr. ***dispar*** and ***A. ulmi***.

Fertilized females of ***A. fraxiniflora*** overwinter under bud scales or in bark crevices. They move to developing male flowers in the spring, and their feeding stimulates gall formation. Eggs are laid in developing galls. After hatching, nymphs pass through two instars before becoming adults. Several generations occur throughout the summer. A generation of overwintering females is produced in the fall. These become fertilized and move to overwintering sites. The life cycle of many ***Eriophyes*** species is similar to that of *A. fraxiniflora*.

Damage

Infestations of ***A. parapopuli*** are most severe on hybrid poplars in the southern prairie provinces. Infested trees lose aesthetic value quickly because new galls are formed every year and old ones persist for several years. A persistent infestation also causes lower branches to become stunted, often resulting in branch death. Trees are seldom killed, but continued heavy attack may increase their susceptibility to other pests, frost, and drought. The other ***Aceria*** and ***Eriophyes*** species cause little damage to their hosts but reduce the aesthetic value of trees in ornamental settings.

Figure 76. Mite galls. A. Adults of *Aceria*. **B.** Galls on aspen caused by *A.* nr. *dispar*. **C.** Gall on poplar caused by *A. parapopuli*. **D.** Poplar twig heavily infested by *A. parapopuli*. **E.** Galls on ash caused by *A. fraxiniflora*. **F.** Pocket galls on elm leaves caused by *A. ulmi*. **G.** Bladder galls on aspen caused by *Eriophyes* species.

A
B
C
D
E
F
G
H

Symptoms and Signs

Many insects and mites cause galls on hardwood foliage and buds. The shape, color, and location of galls, and the tree host species help identify the cause. Galls caused by ***Pemphigus*** aphids (Homoptera: Aphididae) are common on poplars (*Populus* spp.) and aspen (*P. tremuloides* Michx.). Species causing galls on petioles are the lettuce root aphid, ***P. bursarius*** (L.), which causes sickle-shaped galls (A, B); the poplar petiole gall aphid, ***P. populitransversus*** Riley, which causes galls at the base of petioles (E); and a ***Pemphigus*** sp. (D) that causes irregularly shaped galls. Other species such as ***P. populivenae*** Fitch cause galls on poplar and aspen leaf blades (F). Numerous other species of ***Pemphigus*** cause similar galls, but much taxonomic research is required before many of these species can be identified. Aphids feeding inside galls are soft-bodied, 1.0–1.5 mm long, pale yellow to grayish, and often covered by whitish woolly flocculence (C). The poplar vagabond aphid, ***Mordvilkoja vagabunda*** (Walsh) (Aphididae), causes large, irregularly shaped galls on the terminal shoots of poplars (G, H). Young galls are light green (G) but turn reddish brown (H) as they age. These galls remain on the trees after the leaves have fallen. Aphids are greenish to creamy-white, and 4–5 mm long when mature.

Distribution and Hosts

The genus ***Pemphigus*** occurs throughout North America but the distribution of many of the gall-producing species is not known because of the taxonomic uncertainty within the genus. Many species have alternating generations on different hosts (often vegetable crops) or different parts of the same host. All species of poplar are attacked by at least one species of *Pemphigus*. ***Mordvilkoja vagabunda*** occurs throughout most of the northern United States and southern Canada, east of the Rocky Mountains. It attacks numerous poplar species but has been particularly abundant on aspen and balsam poplar (*Populus balsamifera* L.) in the prairie provinces. The secondary host of the poplar vagabond aphid is loosestrife (*Lysimachia* spp.).

Life Cycle

All gall-producing *Pemphigus* are believed to have similar life cycles. ***Pemphigus bursarius*** will be used as an example. Overwintering eggs on poplar twigs hatch in the spring to produce female nymphs. Nymphs move onto petioles, where their feeding stimulates gall formation. They develop into wingless, parthenogenetic females; these give birth to nymphs that develop in the galls and become winged parthenogenetic females. These females disperse to lettuce plants, where several parthenogenetic generations are produced (and feed) on lettuce roots. Winged females produced in the fall fly back to poplars, where they give birth to small, mouthless aphids. These develop into tiny males and females. Mated females each produce a single egg, which overwinters.

Eggs of ***M. vagabunda*** overwinter in or near galls from the previous year. In the spring, nymphs feed in groups on developing shoots, which stimulates gall formation. From this point the life cycle is similar to that described above, with loosestrife serving as the secondary host. A generation of winged males and females is produced on loosestrife in the fall. Mated females lay eggs on poplar.

Damage

These species cause little damage to their hosts, but severe infestations make trees unsightly.

Figure 77. Aphid galls on poplars and aspen. A–C. *Pemphigus bursarius.* **A.** Young gall on poplar. **B.** Mature galls on poplar. **C.** Opened gall showing aphids inside. **D.** *Pemphigus* sp. galls on poplar petioles. **E.** Gall on a cottonwood leaf petiole caused by *P. populitransversus.* **F.** Gall on balsam poplar leaf caused by *P. populivenae.* **G.** New gall on balsam poplar caused by *Mordvilkoja vagabunda.* **H.** Old *M. vagabunda* gall.

A
B
C
D
E
F
G

Symptoms and Signs

Many insects and mites cause galls on hardwood foliage and buds. The shape, color, and location of galls, and the host tree species can help identify the cause. The willow redgall sawfly, ***Pontania proxima*** (Lepeletier) (Hymenoptera: Tenthredinidae), produces reddish brown, bean-shaped galls up to 12 mm in diameter (B). A row of galls usually occurs on both sides of the midrib. Larvae inside galls are pale green with black heads (A). Another ***Pontania*** species produces single spherical galls on sandbar willow (*Salix exigua* Nutt.) leaves. A gall wasp, ***Callirhytis*** nr. ***flavipes*** (Gillette) (Hymenoptera: Cynipidae), alternately produces leaf and twig galls. Pale green galls on leaves occur along the midrib (E). The overwintering generation causes inconspicuous galls on twigs in late July and August. These twig galls are usually not detected until woodpeckers cause damage by foraging for larvae in the galls in winter (F). Adults emerge from twig galls in the spring by chewing small exit holes through the bark (G). Larvae in twig and leaf galls are 2–3 mm long, legless, and have yellowish bodies and head capsules (D).

Distribution and Hosts

Pontania proxima was introduced into North America from Europe and is distributed from eastern Canada to Alberta. It attacks various willows (*Salix* spp.). The distribution of ***C.*** nr. ***flavipes*** is not known, but it appears to be found throughout the prairie provinces, where it attacks only bur oak (*Quercus macrocarpa* Michx.).

Life Cycle

Pontania proxima has one generation per year and overwinters as prepupae in the duff. The black, 4- to 6-mm-long adult sawflies emerge in late spring. Eggs deposited in slits cut into the leaf tissue initiate gall formation, but larval feeding further stimulates gall growth. In late summer mature larvae chew an exit hole through the galls and enter the duff to spin cocoons, in which they overwinter.

Callirhytis nr. ***flavipes*** has two generations per year. Adults of the overwintering generation emerge in late spring. All of the adults are females that reproduce parthenogenetically. They are minute, dark brown or black wasps, about 2.5 mm long, with pale legs and antennae (C). They are believed to enter the buds and lay eggs on the midribs of the rudimentary leaves. The eggs and subsequent feeding of the larvae stimulate the formation of single- and multi-celled galls on the affected leaves by early summer. Pupation occurs in the galls in early July and adults of both sexes emerge in mid-July. Mated females lay eggs in the bark of twigs. This stimulates the formation of single-celled galls, each containing a larva, which overwinters. Pupation occurs in the spring.

Damage

Severe infestations of ***P. proxima*** or of the summer generation of ***C.*** nr. ***flavipes*** appear to cause little serious damage to hosts; however, the conspicuous galls reduce the aesthetic value of ornamental trees. The overwintering generation of ***C.*** nr. ***flavipes*** may kill young twigs. This produces an undesirable bushy growth that detracts from the tree's appearance. The greatest amount of damage is caused by downy woodpeckers looking for larvae. They sometimes remove most of the outer bark from the branches and upper trunk of young trees (F).

Figure 78. Galls on willow and oak. A, B. *Pontania proxima*. **A.** Larva in gall. **B.** Galls on willow leaf. **C–G.** *Callirhytis* nr. *flavipes*. **C.** Adult. **D.** Overwintering larva. **E.** Galls on bur oak leaf caused by the summer generation. **F.** Downy woodpecker damage (see arrows) to bur oak branch containing overwintering larvae. **G.** Bur oak twig showing galls and emergence holes of the overwintering generation.

A
B
C
D
E
F

Symptoms and Signs

There are some conspicuous symptoms for each leaf spot disease (A–F), but microscope examination of spores (mostly conidia) is required for accurate diagnosis of most species. The characteristic symptoms and signs of each leaf spot disease are described in Table 6.

Cause

Ten common or conspicuous leaf spot diseases of aspen and poplar (*Populus* spp.) are listed in Table 6. All causal fungi belong to the subdivision Ascomycotina, but many of them are found only as their anamorphs (imperfect states) during the growing season. Poplar leaf rusts and poplar leaf and shoot blight are discussed separately (*see* Leaf Rusts of Aspen and Poplar, and Leaf and Twig Blight of Aspen and Poplar).

Distribution and Hosts

No detailed distribution data are available for poplar leaf spot diseases in the prairie provinces. The main hosts for each disease are listed in Table 6.

Disease Cycle

In general, airborne ascospores are produced on overwintered dead leaves in spring, and they initiate infections on newly formed leaves. Conidia can cause new infections during the growing season.

Damage

Infections are often heavy and cause premature shedding of leaves; however, leaf spot diseases seldom significantly affect the health of trees unless repeated severe infections occur.

Figure 79. Leaf spot diseases of aspen and poplar. A, B. Ink spot, caused by *Ciborinia whetzelii*, on aspen (*Populus tremuloides* Michx.). **A.** A heavily infected tree. **B.** Infected leaves with several dark round areas (pseudosclerotia). **C, D.** A leaf blight caused by *Linospora tetraspora*. **C.** An infected leaf of balsam poplar (*P. balsamifera* L.). **D.** Close-up of an infected leaf showing black fruiting bodies (ascostromata). **E.** Leaves of aspen infected with *Marssonina populi*. **F.** Black dotlike pycnidia of *Mycosphaerella populorum* on an aspen leaf.

Table 6. Common leaf spot diseases of aspen and poplar in the prairie provinces

Organism	Major hosts	Symptoms and conidia
Ciborinia whetzelii (Seaver) Seaver **Ink spot**	Aspen (*Populus tremuloides*)	Black, round sclerotia on discolored leaves drop during suμmer, leaving shot holes; no conidia (Fig. 79A, B)
Linospora tetraspora G. Thompson **Leaf blight**	Balsam poplar (*Populus balsamifera*)	Large brownish gray discolored areas, scattered black stromata on upper surface; no conidia (Fig. 79C, D)
Marssonina balsamiferae Y. Hirat. **Leaf spot**	Balsam poplar	Brown, angular spots on undersurface of leaves; conidia two-celled, 18–21 × 4.5–5.0 μm, tip cells pointed
Marssonina populi (Lib.) Magn. **Leaf spot**	Aspen, balsam poplar	Brownish round spots on upper side of leaves; conidia two-celled, unevenly divided, 19.0–23.5 × 6.5–8.0 μm (Fig. 79E)
Marssonina tremuloides Kleb. **Leaf spot**	Aspen	Brownish angular spots on underside of leaves; conidia two-celled, unevenly divided, 12–17 × 4–6 μm
Mycosphaerella populicola G. Thompson (anam. *Septoria populicola* Peck) **Leaf spot**	Balsam poplar	Round leaf spots; sporulates on both sides of leaves; conidia two- to five-septate, 45–80 × 3.5–4.5 μm
Mycosphaerella populorum G. Thompson (anam. *Septoria musiva* Peck) **Leaf spot**	Balsam poplar, aspen, and hybrid poplars	Round leaf spots; sporulates on both sides of leaves; conidia one- to four-septate, 28–54 × 3.5–4.0 μm; also known to cause stem cankers (Fig. 79F)
Septogloeum rhopaloideum Dearn. & Bisby **Leaf spot**	Aspen	Large discolored areas, small blisterlike pustules on underside of leaves; conidia usually three-celled, 40–50 × 8–12 μm
Venturia macularis (Fr.) Mueller & Arx (anam. *Pollaccia radiosa* (Lib.) Baldacci & Cif.) **Leaf and twig blight** (*see* Leaf and Twig Blight of Aspen and Poplar)	Aspen	Black shoot blight and leaf necrosis; conidia mostly two-celled, 18–26 × 5–8 μm (Fig. 80A, B)
Venturia populina (Vuill.) Fabric. (anam. *Pollaccia elegans* Servazzi) **Leaf and twig blight** (*see* Leaf and Twig Blight of Aspen and Poplar)	Balsam poplar	Black shoot blight and leaf necrosis; conidia mostly two-celled, 24.5–30.5 × 9–11 μm (Fig. 80C)

A

B

C

Symptoms and Signs

Blackening and wilting of young shoots and leaves are typical symptoms of these two diseases (A–C). Tips of the blackened shoots often bend back, producing the shepherd's crook symptom (B, C). On older leaves, brownish black, irregularly shaped spots appear (A). Typical conidia produced on infected leaves and young shoots permit identification of the diseases (Table 6).

Cause

Two very similar leaf and twig blight diseases are caused by closely related fungal pathogens: ***Venturia macularis*** (Fr.) Mueller & Arx (= *V. tremulae* Aderh., anamorph *Pollaccia radiosa* (Lib.) Baldacci & Cif.) on aspen (*Populus tremuloides* Michx.) and ***V. populina*** (Vuill.) Fabric. (anamorph *Pollaccia elegans* Servazzi) (Ascomycotina: Dothideales) on balsam poplar (*P. balsamifera* L.).

Distribution and Hosts

The two diseases are distributed widely throughout the prairie provinces and other parts of Canada. ***Venturia macularis*** has also been observed on narrowleaf cottonwood (*P. angustifolia* James) and largetooth aspen (*P. grandidentata* Michx.). The main host of ***V. populina*** is balsam poplar, but black cottonwood (*P. trichocarpa* Torr. & Gray) is also known to be a host.

Disease Cycle

Ascospores and conidia produced on dead overwintering tissues are dispersed by wind and infect young shoots and leaves in the spring. Conidia are produced on the blackened part of stems and leaves, and reinfect young leaves. Once leaves develop fully and harden, no new infections seem to occur. Later in the fall and during the winter, the perfect stage of the fungi (perithecia) begins to form on dead infected tissues.

Damage

When most of the tender shoots of young trees are attacked, the trees are disfigured and growth is severely affected. Infection on larger trees is rare and does not significantly affect their growth. These two diseases are expected to become increasingly important as aspen and poplar are managed intensively in short rotations for pulp and lumber or for biomass to be used for energy.

Figure 80. Leaf and twig blight of aspen and poplar. A, B. *Venturia macularis* on aspen. **A.** Infection on a young leaf. **B.** Shepherd's crook symptom on a young shoot. **C.** *V. populina* on balsam poplar, showing shepherd's crook symptom on a young shoot.

A
B
C
182

Symptoms and Signs

The uredinial state of poplar leaf rusts appears as powdery, golden-yellow pustules (B) on the underside of leaves and as yellow spots on the upper side of leaves. The two species can be distinguished by the size of urediniospores (15–23 × 23–35 µm for ***Melampsora albertensis*** Arthur and 16–29 × 32–45 µm for ***M. occidentalis*** Jacks.). Later in the summer the telial state appears among uredinia as smooth, raised, orange-yellow pustules (B). Telia turn from dark brown to black toward the end of the growing season (D). Heavily infected leaves are paler in color and tend to drop prematurely.

Cause

Two species of rust fungi (Basidiomycotina: Uredinales) are known to cause leaf rust on aspen (*Populus tremuloides* Michx.) and poplar (*Populus* spp.) in the prairie provinces. They are ***Melampsora albertensis*** (A, B) and ***M. occidentalis*** (C). ***Melampsora albertensis*** has often been called *M. medusae* Thuemen, but true ***M. medusae*** is an eastern North American species, and in the prairie provinces it occurs only sporadically in eastern Manitoba.

Distribution and Hosts

Melampsora albertensis occurs on aspen in western North America. Distribution of ***M. occidentalis*** is also restricted to western North America, and in the prairie provinces it occurs in Alberta and Saskatchewan, commonly on balsam poplar (*P. balsamifera* L.), black cottonwood (*P. trichocarpa* Torr. & Gray), and hybrid poplars. Several species of larch (*Larix* spp.) are native alternate hosts of these two rusts in the prairie provinces, but other conifers, such as Douglas-fir (*Pseudotsuga menziesii* (Mirb.) Franco), pines (*Pinus* spp.), and spruces (*Picea* spp.), are known to be alternate hosts for these rusts elsewhere.

Disease Cycle

The powdery, yellowish orange spores, urediniospores, produced on poplars can reinfect poplar leaves repeatedly during the growing season, thus intensifying the infection. Later in the season, teliospores are produced on the leaves. Teliospores overwinter on dead fallen leaves and germinate in the spring to produce basidiospores, which infect coniferous alternate hosts, on which they produce spermogonia and aecia. Aeciospores produced in aecia on conifer needles infect poplar leaves in early summer. There are strong indications that the two rusts can overwinter on poplar and start infection in the spring without going to an alternate host.

Damage

Because heavily infected leaves defoliate early, the general health of the trees can be affected; however, these rusts seldom kill infected trees. The damage becomes significant only in situations of intense cultivation and management or on poplar trees of ornamental value.

Figure 81. Leaf rusts of aspen and poplar. A, B. *Melampsora albertensis*. **A.** Two infected aspen leaves. **B.** Close-up showing powdery, yellow uredinia and dark yellow to brown raised pustules of telia. **C.** *M. occidentalis* infection on balsam poplar.

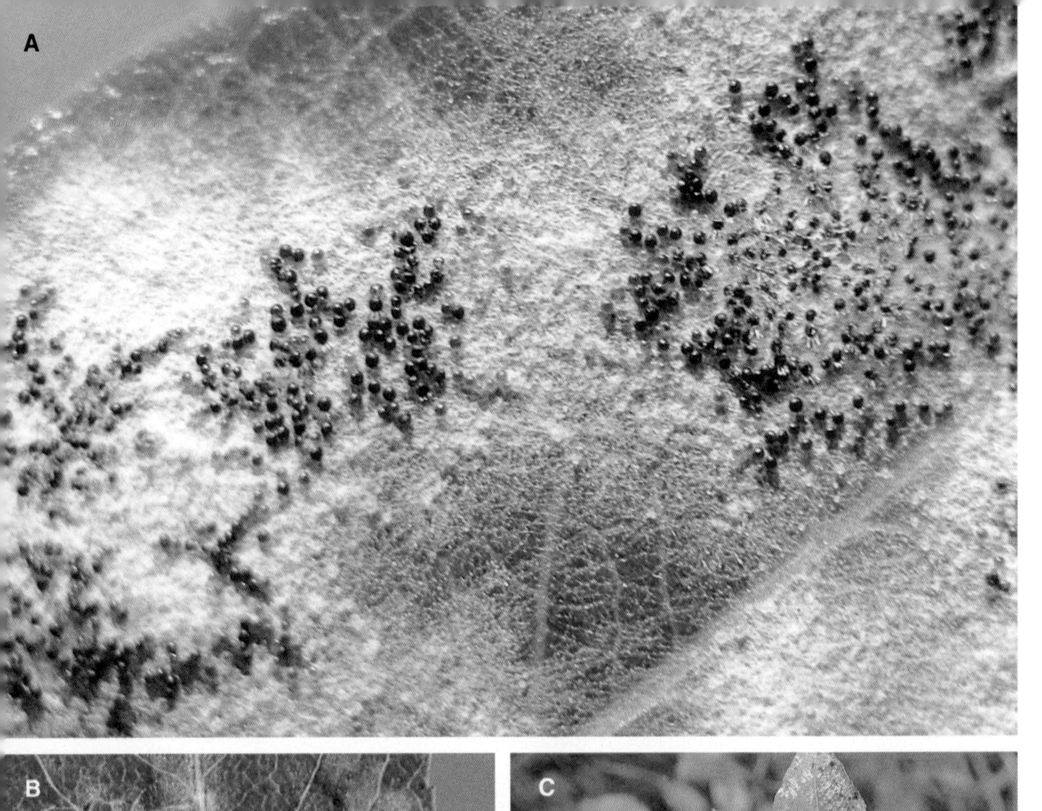
A

B
184

C

Symptoms and Signs

Powdery mildews are characterized by the presence of grayish white, powdery or dusty fungal growth on the surface of leaves and production of small brown to black spherical fruiting bodies (cleistothecia) (A–C). Heavily infected leaves may be discolored or disfigured, but light infection does not affect the appearance of leaves. Microscopic characteristics have to be examined to identify the pathogens to the species.

Cause

Four species of powdery mildew fungi are common on various broadleaf trees in the prairie provinces. They are ***Erysiphe aggregata*** (Peck) Farlow, ***Microsphaera penicillata*** (Wallr.:Fr.) Lév., ***Phyllactinia guttata*** (Wallr.:Fr.) Lév., and ***Uncinula adunca*** (Wallr.:Fr.) Lév. (= *U. salicis* (DC.) Wint.) (Ascomycotina: Erysiphales).

Distribution and Hosts

All four powdery mildew species have a wide host range, including many common hardwood trees and herbaceous plants in the prairie provinces. Major tree hosts for each pathogen are listed in Table 7. These fungi do not occur on conifers.

Disease Cycle

Powdery, white conidia are produced on the surface mycelium and can reinfect the host during the growing season. Cleistothecia (A), which are produced later in the growing season, overwinter on dead fallen leaves. Ascospores produced in the cleistothecia are released and infect newly formed leaves in the spring.

Damage

Powdery mildews do not kill their hosts; however, heavy infection can affect the general health and appearance of trees and thereby reduce their horticultural value.

Table 7. Powdery mildew species common on broadleaf trees in the prairie provinces

Fungus	Hosts
Erysiphe aggregata (Peck) Farlow	Alder (*Alnus* spp.)
Microsphaera penicillata (Wallr.:Fr.) Lév.	Alder, lilac (*Syringa vulgaris* L.), maple (*Acer* spp.)
Phyllactinia guttata (Wallr.:Fr.) Lév.	Alder, apple and crab apple (*Malus* spp.), cherry (*Prunus* spp.), mountain ash (*Sorbus* spp.), pear (*Pyrus* spp.), saskatoon (*Amelanchier alnifolia* Nutt.), willow (*Salix* spp.)
Uncinula adunca (Wallr.:Fr.) Lév.	Aspen and poplar (*Populus* spp.), willow (A–C)

Figure 82. Powdery mildew of poplar and willow, caused by *Uncinula adunca*. A. Close-up of infected leaf surface of aspen (*Populus tremuloides* Michx.) showing black (mature) and pale yellow (immature) cleistothecia. **B.** An infected willow leaf (*Salix* sp.). **C.** Heavily infected aspen leaves.

A
B
C
D
E
F

LEAF AND BERRY RUSTS OF SASKATOON AND OTHER ROSACEOUS HOSTS

Symptoms and Signs

Symptoms and signs distinguishing the four rust fungi causing leaf and berry rusts of saskatoon (*Amelanchier alnifolia* Nutt.) and other common rosaceous hosts in the prairie provinces are compared and listed in Table 8.

Cause

Four species of rust fungi belonging to the genus ***Gymnosporangium*** (Basidiomycotina: Uredinales) attack saskatoon and other rosaceous hosts in the prairie provinces (Table 8).

Distribution and Hosts

The four rust species are widely distributed in all prairie provinces. ***Gymnosporangium clavariiforme*** (Pers.) DC. (A, B), ***G. clavipes*** Cooke & Peck (C, D), and ***G. nidus-avis*** Thaxter parasitize many rosaceous hosts, including saskatoon, but ***G. nelsonii*** Arthur (E, F) is apparently found only on saskatoon. Alternate hosts (telial hosts) of all four rusts include various species of juniper (*Juniperus* spp.) (Table 8).

Disease Cycle

The basic life cycle of the four rust species is the same. Teliospores are produced on perennial infections of juniper. In the spring or early summer, they germinate and produce basidiospores, which infect saskatoon (and other rosaceous hosts) and produce spermogonia and aecia. Aeciospores produced in aecia cause infection on juniper. There is no uredinial stage in these rusts.

Damage

Of the four species, ***G. clavipes*** and ***G. nelsonii*** are the most common and cause significant economic damage, affecting small fruit production.

Figure 83. Leaf and berry rusts of saskatoon. A, B. *Gymnosporangium clavariiforme.* **A.** Leaf and fruit infections with aecia on saskatoon. **B.** Telial horns on common juniper (*Juniperus communis* L.). **C, D.** *G. clavipes.* **C.** Infected fruits of saskatoon showing aecia. **D.** Raised reddish brown telia on a stem of common juniper. **E, F.** *G. nelsonii.* **E.** Leaf infections on saskatoon showing dark pointed aecia. **F.** Telial galls with gelatinous telial horns on creeping juniper (*J. horizontalis* Moench.). Unhydrated (right) and hydrated (left) galls.

Table 8. Symptoms, signs, and alternate hosts of the four species of *Gymnosporangium* that attack saskatoon in the prairie provinces

Species of *Gymnosporangium*	Telial hosts	Symptoms and signs on saskatoon
G. clavariiforme (Pers.) DC.	*Juniperus communis* L. var. *depressa* Pursh, on fusiform swellings (B)	Mostly on fruits and stems; aecia cup dehisces from the apex, cinnamon-brown; aeciospores 23–36 (up to 41) × 21–32 μm; wall yellow-brown, verrucose; germ pores scattered, 8–10 (A)
G. clavipes Cooke & Peck	*J. communis* var. *depressa*, on slight fusiform swellings (D)	Mostly on fruits but also stems; aecia cup shaped, orange-yellow; aeciospores 28–46 × 22–38 μm, contents bright orange; wall colorless, densely verrucose; germ pores obscure, about 8 (C)
G. nelsonii Arthur	*J. horizontalis* Moench and *J. scopulorum* Sarg., on globose galls up to 2–3 cm (F)	Mostly on undersides of leaves but also on fruits; aecia dehisce both from apex and by lateral slits, chocolate-brown; aeciospores 20–32 × 17.5–28.0 μm; wall yellow-brown, densely verrucose; germ pores scattered, 6–10 (E)
G. nidus-avis Thaxter	*J. horizontalis* and *J. scopulorum*, on needles, causing witches' brooms	Mainly on underside of leaves but also on stems or fruits; aecia dehisce by lateral slits, brown; aeciospores 20.0–33.5 × 18.5–25.0 μm; wall dark yellow-brown, densely verrucose; germ pores scattered, 8–10

A

B

C

BLACK LEAF AND WITCHES' BROOM OF SASKATOON

Symptoms and Signs

Undersides of infected leaves are covered by an olive-brown to black fungal mat (A, B), which produces conidia and develops numerous black globose fruiting bodies (pseudothecia) that contain ascospores. Infected shoots form witches' brooms (C), which enlarge each year. Dead blackened leaves remain on the brooms during the winter and can be spotted easily.

Cause

Black leaf and witches' broom of saskatoon (*Amelanchier alnifolia* Nutt.) is caused by a fungus, ***Apiosporina collinsii*** (Schw.) Hoehnel (≡ *Dimerosporium collinsii* (Schw.) Thuemen) (Ascomycotina: Dothideales). A closely related fungus, ***A. morbosa*** (Schw.) Arx, causes black knot disease of choke cherry and pin cherry (*see* Black Knot of Cherry).

Distribution and Hosts

This disease is common throughout western North America on saskatoon but is known on several other species of *Amelanchier* in other regions.

Disease Cycle

The fungus overwinters on dead, infected leaves and in the spring produces two kinds of infection spores (ascospores and conidia), which infect new shoots. Infections are systemic and perennial; therefore new leaves produced on previously infected shoots are infected.

Damage

The growth of heavily infected plants is greatly reduced, and if all parts of a plant are infected the tree dies in 1 or 2 years. Heavy infections reduce aesthetic value of infected trees.

Figure 84. Black leaf and witches' broom of saskatoon, caused by *Apiosporina collinsii*. A. An infected plant showing the black undersides of leaves. **B.** Close-up of the underside of an infected leaf. **C.** Witches' broom symptom of an infected plant.

A

B

C

Symptoms and Signs

The first noticeable symptom is a silvering or leaden luster of the leaves (hence the name silver leaf), which is followed by browning of the leaf midribs and the leaf margins (A). Affected parts of the tree gradually lose vigor and eventually die (B). If the main stem is infected, the tree usually dies within 2–3 years. Small shelflike fruiting bodies with a grayish white, hairy or velvety upper surface and purplish, smooth undersurface are produced at the bottom of affected trees (C). They are usually produced on pruning scars, branch stubs, or cracks on stems.

Cause

Silver leaf disease is caused by a fungus, ***Chondrostereum purpureum*** (Pers.:Fr.) Pouzar (≡ *Stereum purpureum* (Pers.:Fr.) Fr.) (Basidiomycotina: Aphyllophorales).

Distribution and Hosts

Silver leaf disease occurs widely in the world, especially in urban and cultivated conditions. The causal organism has a wide host range. Major tree hosts in the prairie provinces are mountain ash (*Sorbus* spp.), apple (*Malus pumila* (L.) Mill.), hawthorn (*Crataegus* spp.), crab apple (*Malus* spp.), pear (*Pyrus* spp.), plum (*Prunus* spp.), poplar (*Populus* spp.), willow (*Salix* spp.), cotoneaster (*Cotoneaster* spp.), Nanking cherry (*Prunus tomentosa* Thunb.), and spruce (*Picea* spp.).

Disease Cycle

Basidiospores produced on the fruiting bodies infect open wounds or branch stubs and invade the sapwood, where the fungus causes decay. Although the fungus is usually confined to small areas around the initial infection, toxins produced by the organism affect the whole tree. Fruiting bodies are produced on standing trees but are also common on dead stumps and dead logs on the ground.

Damage

This is one of the most common and conspicuous diseases of ornamental and fruit trees in the prairie provinces and is most commonly seen on old, well-established trees. It causes a serious decay and reduces the value and yield of trees.

Figure 85. Silver leaf, caused by *Chondrostereum purpureum*. A. An affected branchlet (left) and a healthy one (right) of mountain ash. **B.** An infected mountain ash tree showing dieback and decline. **C.** Small shelflike fruiting bodies with whitish hairy upper surface and smooth purple undersurface.

A

B

C

Symptoms and Signs

The most conspicuous symptom of tar spot is the presence of circular, raised black mats of fungal tissue (stromata), 2–5 mm in diameter, scattered irregularly on the upper surface of leaves (A, B). Narrow yellowish bands of discoloration usually surround the stromata (B). Two kinds of spores are produced on the stroma. Ascospores are produced on grayish disks (apothecia) produced in the stroma and become exposed by fissures in the crust. Conidia are produced, usually before ascospores, in small dotlike fruiting structures (pycnidia) embedded in the stroma. The main diagnostic characteristic of the black vein disease is a black, raised ridge of fungal tissue (pseudosclerotium) along disfigured midribs of willow leaves (C). Small (up to 6 mm in diameter) cup-shaped apothecia with stalks are produced in overwintered pseudosclerotia on the ground.

Cause

Tar spot of willow (*Salix* spp.) is caused by ***Rhytisma salicinum*** (Pers.:Fr.) Fr. (Ascomycotina: Rhytismatales), and black vein is caused by ***Ciborinia foliicola*** (Cash & Davids.) Whet. (Ascomycotina: Helotiales).

Distribution and Hosts

The two diseases occur widely on various species of willow in the prairie region, but black vein is less known than tar spot, probably because it has a less conspicuous appearance.

Disease Cycle

New infections of both tar spot and black vein of willow are initiated in the spring by ascospores produced on dead overwintered leaves on the ground. In the tar spot disease, conidia likely function as sexual spores (spermatia) and do not have the ability to initiate new infections.

Damage

Although the two diseases are conspicuous and infection is sometimes very heavy, they do not significantly affect the health of the host tree.

Figure 86. Tar spot and black vein of willow. A, B. *Rhytisma salicinum.* **A.** A willow plant heavily infected with tar spot. **B.** Close-up of an infected leaf showing black, raised stromata. **C.** Curled leaves showing black vein symptom (arrow), caused by *Ciborinia foliicola.*

INSECTS AND DISEASES OF BROADLEAF TREES

Roots, Stems, and Branches

A
B
C
D
E
F

Symptoms and Signs

The first conspicuous sign that trees are infested by the poplar borer, ***Saperda calcarata*** Say (Coleoptera: Cerambycidae), is varnishlike resin flowing down the stems and staining the bark (E). This resin comes from wounds in the bark (C) made by feeding larvae. Some of these holes have boring dust exuding from them (D). If bark is stripped from the infested trees, larvae and their galleries are observed in the bark and wood. Larvae are legless grubs with creamy-white bodies, brown heads and thoracic shields, and are about 40 mm long when fully grown (B). Bark of trees previously infested by poplar borers is marked by crevices and fissures as a result of larval feeding and woodpecker damage (F).

Distribution and Hosts

The poplar borer has a transcontinental distribution. It attacks primarily trembling aspen (*Populus tremuloides* Michx.), but also other poplars (*Populus* spp.) and willows (*Salix* spp.).

Life Cycle

The poplar borer requires 3–5 years (typically 4 years) to complete development, depending on climate; however, populations are not synchronized so adults are produced each year. Adult beetles (A) are about 25 mm long. The body is basically gray but the dorsal surface has faint yellow stripes on the thorax and yellow blotches on the elytra, all densely stippled with small brown dots. Adults emerge in late June and July and live for up to 6 weeks. They feed on aspen and willow foliage and begin laying eggs about 1 week after emergence. Females cut crescent-shaped notches into the bark and deposit one or two eggs in each notch. Oviposition tends to be concentrated on sun-exposed parts of the trunk or in the lower crown. Eggs hatch in about 3 weeks and young larvae feed on the inner bark until about October, when they hibernate. Larvae begin feeding again in April or May, when they bore through the sapwood into the heartwood, often ejecting sawdust (D) from the surface wound on the tree. The second winter is spent in a cell formed from tightly packed frass at the distal end of the burrows. Third-year feeding also begins in April or May. Larvae cease feeding in August and construct hibernation cells at the distal end of burrows. The insects spend the third winter as prepupae. Pupation starts by mid-May or early June of the fourth year, and adults emerge soon afterward.

Damage

Open-growing trees, and those around the edges of stands seem to be most vulnerable to attack. This insect is particularly troublesome in aspen parkland, where up to 75% of trees 7–10 cm in diameter may be infested. This species also occurs in forested areas but is usually not a serious problem. The same trees may be repeatedly attacked by poplar borers. Trees are usually not killed by poplar borer attack, even when riddled with tunnels, but the weakened stems are liable to break during windstorms, and the wood is almost useless for lumber or other purposes. Woodpeckers cause much damage to wood while searching for larvae (F), and the openings maintained by larvae for ejection of boring dust provide infection courts for various rot fungi.

Figure 87. Poplar borer, *Saperda calcarata*. A. Adult. **B.** Larva. **C.** Wound in the bark caused by larva, and associated sap flow. **D.** Sawdust ejected by larva. **E.** Fresh damage to trees (note sap flow). **F.** Old poplar borer and woodpecker damage.

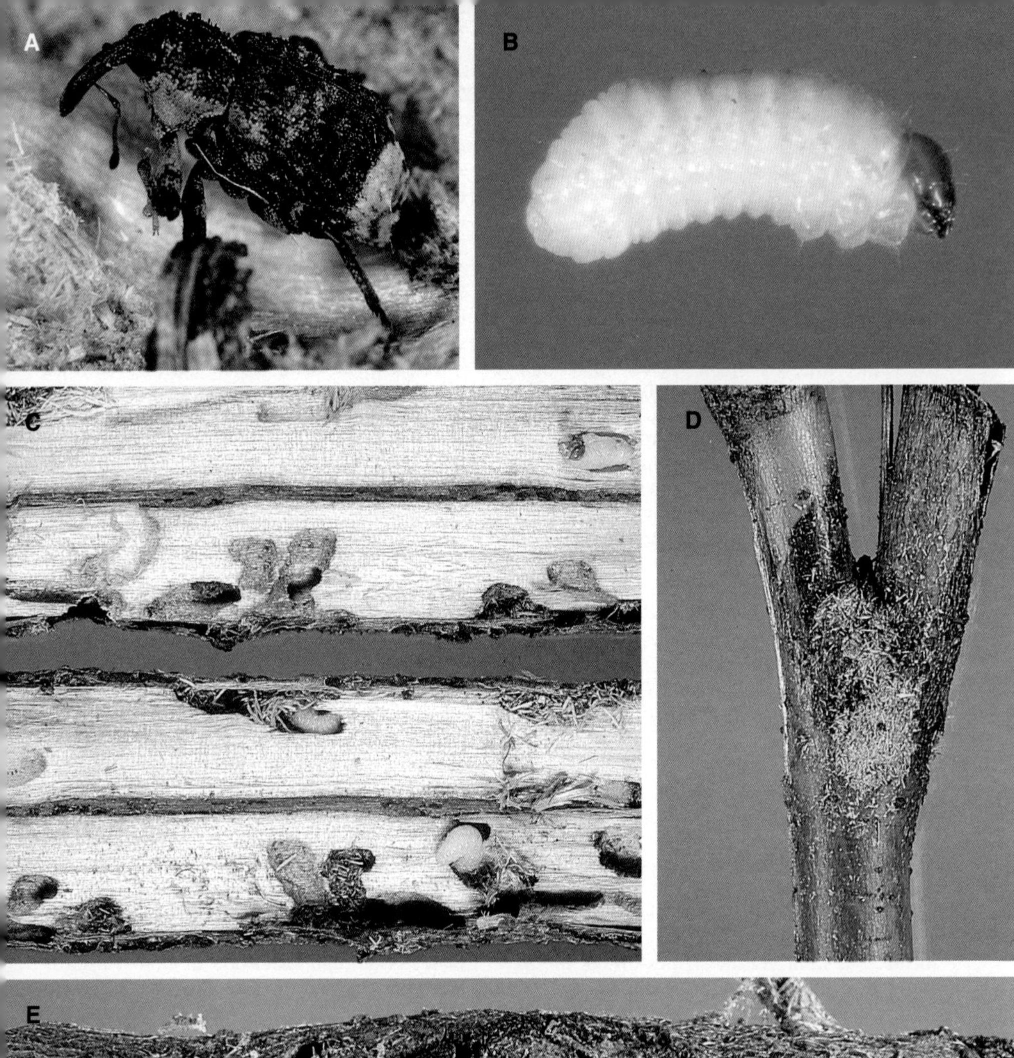
A
B
C
D

E

Symptoms and Signs

Current infestations of poplar and willow borer, ***Cryptorhynchus lapathi*** (L.) (Coleoptera: Curculionidae), are recognized by irregular splits and holes in the bark of saplings, through which moist, reddish brown boring dust and sap exude (D). If infested stems are split open, larvae and their tunnels are visible (C). Larvae are creamy-white, brown-headed, C-shaped, legless grubs that are about 13 mm long when mature (B). Previously infested stems may be easily identified by the presence of circular (3–4 mm diameter) emergence holes chewed by new adults in the bark (E), as well as calluses over injured areas. Dissection of these stems also reveals old larval tunnels.

Distribution and Hosts

This species was introduced from Europe into North America and now has a near-transcontinental distribution in southern Canada and the northern United States. It attacks primarily willow (*Salix* spp.) and poplars (*Populus* spp.), but sometimes alder (*Alnus* spp.) and white birch (*Betula papyrifera* Marsh.).

Life Cycle

This species has a 2-year life cycle; however, because populations are not synchronized, adults are produced every year. Adults are stout weevils with long curved snouts and are 8–10 mm long (A). The body is mostly covered with black scales; however, there is an irregular band of pinkish or creamy-white scales across the posterior part of the body. Adults emerge from hibernation sites in the spring and feed on young shoots by inserting the mouthparts into the bark to feed on the phloem. Eggs are laid in the summer in holes chewed by the females in the bark on the lower part of the stems. Larvae burrow in the bark at first, then enter the wood (C), where they excavate long meandering tunnels, pushing the boring dust to the outside of the stem (D). Small larvae overwinter in their tunnels and recommence feeding in the spring. They pupate in the wood in late summer. It is not known if emerging adults can reproduce before overwintering. Adults overwinter in the duff.

Damage

Injury caused by this species is greatest on nursery stock and on newly planted trees in ornamental settings. Larval feeding may directly kill young saplings, but more often weakens the stems of willows and poplars, making them susceptible to breakage during high winds or ice storms. Infested trees may also become bushy from coppice or adventitious growth.

Figure 88. Poplar and willow borer, *Cryptorhynchus lapathi*. A. Adult. **B.** Larva. **C.** Larvae and damage in willow stem. **D.** Willow stem showing boring dust expelled by feeding larvae. **E.** Emergence holes on willow stem.

A
B
C
D
E
F
G
H

Symptoms and Signs

The bronze birch borer, ***Agrilus anxius*** Gory (Coleoptera: Buprestidae), usually attacks declining trees. Therefore trees showing initial signs of decline, that is, dieback of the uppermost twigs and branches, are at high risk. This insect may attack and lay its eggs over the entire main stem and at the base of large branches. Early signs of the presence of larvae may include chlorotic leaves and sparse foliage in the upper crown. Larvae do not expel frass or sap, so bark must be removed from the tree to confirm their presence. Larvae make long meandering, frass-packed galleries in the phloem and cambium (C). Feeding tunnels constructed in healthy tissues are usually unsuccessful and subsequently heal over with callus tissue, which gives the bark a bumpy appearance that remains for many years (D). The bronze poplar borer, ***A. liragus*** Barter & Brown, also attacks weakened trees. There are no obvious external signs of larval galleries, which are similar in shape and size (G) to those of *A. anxius*. Successful attacks by both species are indicated by the presence of D-shaped, 3- to 4-mm-diameter adult emergence holes in the bark on dying or dead portions of stems and branches (E). Larvae of both species are flattened, creamy-white, legless grubs with light brown heads, distinctly segmented bodies, laterally expanded thoraxes, and two dark brown dorsal spines on the posterior part of the body (B, F). Mature larvae are about 35 mm long.

Distribution and Hosts

Both species have transcontinental distributions. ***Agrilus anxius*** attacks mainly native and European white birches (*Betula* spp.), but other birch species are also attacked. ***Agrilus liragus*** attacks trembling aspen (*Populus tremuloides* Michx.) and other poplars (*Populus* spp.).

Life Cycle

Agrilus anxius adults are active from late June to August. They are olive-green to black with metallic-bronze reflections, slender, and 6–11 mm long (A). They feed on birch, aspen, and alder (*Alnus* spp.) leaves for a while, but cause no significant injury. Adults are strong fliers. Mated females lay eggs in bark crevices on weakened or dying trees, usually on the sunny side of open-growing trees. Larvae burrow into the cambial area and excavate meandering tunnels (C) in the phloem, entering the wood periodically to molt. Larvae develop through five instars and overwinter in their tunnels. Pupation occurs in the spring in the larval tunnels and emerging adults chew their way through the bark. ***Agrilus liragus*** adults emerge in early June and feed on aspen leaves before reproducing. The life cycle is similar to that of *A. anxius*.

Damage

Both species are considered to be secondary pests because they mostly injure trees under stress from other factors, especially drought and insect defoliation. Larval tunnels of ***A. anxius*** partially girdle stems and branches and further contribute to decline and death of trees (H). The degree of injury relates to the density and distribution of successful larval galleries, as well as the condition of the host trees. Attack injury has mostly been associated with older, well-established trees beyond the sapling stage. Beetle populations are also known to build up on stressed birches and then move on to attack apparently healthy trees. Damage caused by this species is particularly serious in ornamental settings. The extent of damage caused by ***A. liragus*** is poorly known.

Figure 89. Bronze birch borer and bronze poplar borer. A–E. *Agrilus anxius*. **A.** Adult. **B.** Larva. **C.** Larval gallery. **D.** Bumps on birch branch caused by phloem growth over larval galleries. **E.** D-shaped adult emergence holes. **F.** *A. liragus* larva. **G.** *A. liragus* larval galleries. **H.** Birch dying from *A. anxius* attacks.

A
B
C
D
E
F
G

Symptoms and Signs

The first signs of attack by the ash borer, ***Podosesia syringae*** (Harris) (Lepidoptera: Sesiidae), on stems and branches is the presence of 4- to 8-mm-diameter holes in the bark surface (D) and yellowish brown boring dust on the bark scales under the attack sites and around the base of the tree. Splitting of branches or stems reveals larval feeding tunnels and larvae in the bark and wood. Tunnels are up to 32 mm long and usually darkly stained by fungi (C). Galleries occur throughout the wood of small trees, but only in the sapwood of large trees. Trees attacked for several years develop roughened bark marked with numerous ridges and fissures (D). Larvae are creamy-white with light brown heads and thoracic shields, have reduced thoracic legs and five pairs of abdominal prolegs, and are 26–34 mm long when fully grown (B). Damage by the cottonwood crown borer, ***Sesia tibialis*** (Harris) (Sesiidae), is limited to the lower boles and roots of trees. The first sign of attack by this pest is the presence of reddish brown boring dust around the base of trees. Close examination of the trunk reveals 4- to 8-mm-diameter holes in the bark surface. If the wood is split open, larvae and darkly stained larval feeding galleries (F) are seen. Larvae are similar to those of *P. syringae*, but slightly larger.

Distribution and Hosts

Podosesia syringae is distributed throughout southern Canada and the United States, east of the Rocky Mountains. It attacks green ash (*Fraxinus pennsylvanica* var. *subintegerrima* (Vahl) Fern.) and lilac (*Syringa* spp.). ***Sesia tibialis*** has a transcontinental distribution. It attacks poplar (*Populus* spp.) and willow (*Salix* spp.).

Life Cycle

Podosesia syringae requires 2 years to complete its life cycle in the prairie provinces. Adults emerge in June and early July. The moths have transparent wings, dark brown or black bodies with yellow bands on the abdomen, and a wingspan of about 30 mm (A). Females lay eggs in bark crevices or wounds, usually on the sunny side of open-growing trees. Larvae feed in the bark until fall, then overwinter. They resume feeding in the spring and enter the wood, where they construct galleries up to 32 cm long. After overwintering a second time, larvae bore towards the surface in the spring of the third year, leaving only a thin layer of bark, where they pupate. Pupae push their way to the surface just before adult emergence.

Sesia tibialis has a life cycle similar to that of *P. syringae*. Adults are brownish black moths with yellow markings on the body, transparent wings, and a wingspan of 30–40 mm (E). Larvae living in the stem pupate under the bark; however, larvae in the roots move to the soil to pupate.

Damage

In general, trees that are stressed or have received mechanical damage to the bark are more susceptible to attacks by clearwing moths than are healthy trees. Tunnels made by ***P. syringae*** larvae weaken the stem and allow moisture and fungi to enter, causing further tree decline. Extensive tunnelling in the wood may result in tree breakage in high winds or ice storms. Prolonged attacks result in dead branches and tops, and often tree death, especially during drought. ***Sesia tibialis*** weakens hybrid poplars in stooling beds (G), greatly reducing the yield of cuttings.

Figure 90. Clearwing moths. A–D. *Podosesia syringae*. **A.** Adult. **B.** Larva and gallery in green ash. **C.** Larval galleries and frass in green ash stem. **D.** Heavily infested green ash stem. Note holes in wood and sloughing of bark. **E–G.** *Sesia tibialis*. **E.** Adult. **F.** Larval damage in poplar stem. **G.** Hybrid poplar stool weakened by feeding larvae.

A
B
C
D
E
F

Symptoms and Signs

Larvae of the smaller willowshoot sawfly, ***Euura atra*** (Jurine) (Hymenoptera: Tenthredinidae), feed inside the new shoots of willows. Unfortunately, the presence of larvae inside shoots is difficult to detect early in the season; often the only visible symptom is a slight swelling or roughness of the bark of infested shoots. In early fall, larvae chew exit holes through the bark at one end of their galleries. These holes are generally plugged with bits of frass, wood, and webbing. Externally the exit hole is evident from a slight depressed round area and adjacent discolored bark (C). Shoots exhibiting such symptoms may be split open to reveal larval galleries and overwintering larvae in thin silken cocoons. Larvae have pale green bodies, black heads, thoracic legs, and six pairs of rudimentary abdominal prolegs, and are about 8 mm long when fully grown (D). Without close examination of shoots it is difficult to determine if they are infested until the next spring, when no foliage would appear on the dead shoots (E). During the fall and winter, however, downy woodpeckers often dig out and feed on overwintering larvae; their presence around willows and damage to shoots may be signs of sawfly infestation.

Distribution and Hosts

This species is distributed throughout North America and Europe. It attacks a number of willows (*Salix* spp.), but introduced European willows such as golden willow (*S. alba* L. var. *vitellina* (L.) Stokes) are the preferred hosts in the prairie provinces.

Life Cycle

This species has one generation per year. Larvae overwinter in the mined shoots of the host, usually within thin silken cocoons. Pupation occurs in the shoots in the spring and adults emerge in late May to early June. Adults are 4- to 5-mm-long black sawflies with four membranous wings (A). Using their ovipositors, females cut small holes into the base of shoots and insert eggs (B). After hatching, larvae bore into the pith, where they feed and construct feeding galleries (D). Larvae develop through seven instars by late summer. The frass produced by larvae is pushed to one end of their tunnels; it is not expelled to the surface. In late September or early October, larvae make exit holes through the bark at one end of the tunnels. They then spin silken cocoons at the other end of the tunnels, in which they overwinter.

Damage

Euura atra sometimes causes severe damage to willow in stooling beds and in young shelterbelts (E). Some of this damage, however, is indirect and caused by other agents such as woodpeckers and the fungus ***Cytospora*** sp. (Deuteromycotina: Coelomycetes) (F), associated with the sawflies. Healthy trees can survive repeated attacks, but become bushy and ragged in appearance.

Figure 91. Smaller willowshoot sawfly, *Euura atra*. A. Adult. **B.** Egg. **C.** Emergence hole in shoot and associated discoloration of bark. **D.** Larva in gallery. **E.** Damaged shelterbelt. **F.** The fungus *Cytospora* sp. on willow infested by sawflies.

A
B
C
D
E
F
208

Symptoms and Signs

Four species of scales are most common on hardwoods in the prairie provinces. Their presence is easily determined by close examination of branches and stems at any time of the year. Mature female scales of the European fruit lecanium, ***Parthenolecanium corni*** (Bouché) (Homoptera: Coccidae), are 5–6 mm long, almost hemispherical in shape, and mottled brown in color (C). Males are smaller, whitish, and not as conspicuous. Mature females of the scurfy scale, ***Chionaspis furfura*** (Fitch) (Homoptera: Diaspididae), are pear-shaped, whitish, and 2–3 mm long, and males are smaller and slender (D). Female oystershell scales, ***Lepidosaphes ulmi*** (L.) (Diaspididae), resemble tiny oyster shells. They are about 3 mm long, whitish, and occur in clusters (E). Males are rarely seen. Mature female San José scales, ***Quadraspidiotus perniciosus*** (Comstock) (Diaspididae), are gray, circular, flat with a nipple-like raised area near the center, and 1–2 mm in diameter (F). Males are smaller and oval. When populations of scales are very high, the foliage of infested trees and branches may appear more pale or thinner than that of healthy trees and branches.

Distribution and Hosts

All four species of scales are widespread in North America. ***Parthenolecanium corni*** and ***L. ulmi*** were introduced into North America from Europe and ***Q. perniciosus*** may have originated in China. All species feed on a wide variety of hardwood tree and shrub species, but ***P. corni*** is most common on white elm (*Ulmus americana* L.) and ***C. furfura*** is most common on white elm, ash (*Fraxinus* spp.), and aspen (*Populus tremuloides* Michx.).

Life Cycle

Parthenolecanium corni has one generation per year in the prairie provinces and overwinters as nymphs in bark crevices or under dead female scales attached to the bark (A). Nymphs emerge from hibernation in late March to early May and settle on twigs and branch terminals to feed. Nymphs are whitish, oval, and 1–1.5 mm long (B). Scales feed by inserting their mouthparts into the plant tissue and sucking out the sap. Growth of scales is often completed by mid-May. Winged mature males appear in late May and mate with females. Eggs are laid in June. The powdery, white eggs hatch from late June to late July. The young nymphs migrate to leaves, where they feed throughout the summer. Most of the nymphs congregate on the lower surface of leaves near the larger veins. Nymphs move to their hibernation sites in late summer.

The life cycles of the other three species have not been studied in the prairie provinces. ***Quadraspidiotus perniciosus*** overwinters as nymphs and may have one or two generations per year. ***Chionaspis furfura*** and ***L. ulmi*** are thought to have only one generation per year and overwinter as eggs.

Damage

Light infestations of scale insects cause little damage to hosts, but large infestations may weaken or kill stems and branches. Severe infestations of ***L. ulmi*** have reportedly killed stands of ash in eastern North America, but this is unusual. Damage is usually confined to individual stems and branches on small groups of trees or shrubs, especially those in shaded areas.

Figure 92. Scale insects. A–C. *Parthenolecanium corni.* **A.** Old female scales. One scale has been lifted to show overwintering nymphs underneath. **B.** Nymph. **C.** Mature female scales. **D.** Male (small) and female (large) scales of *Chionaspis furfura.* **E.** Female oystershell scales, *Lepidosaphes ulmi.* **F.** Female San José scales, *Quadraspidiotus perniciosus.*

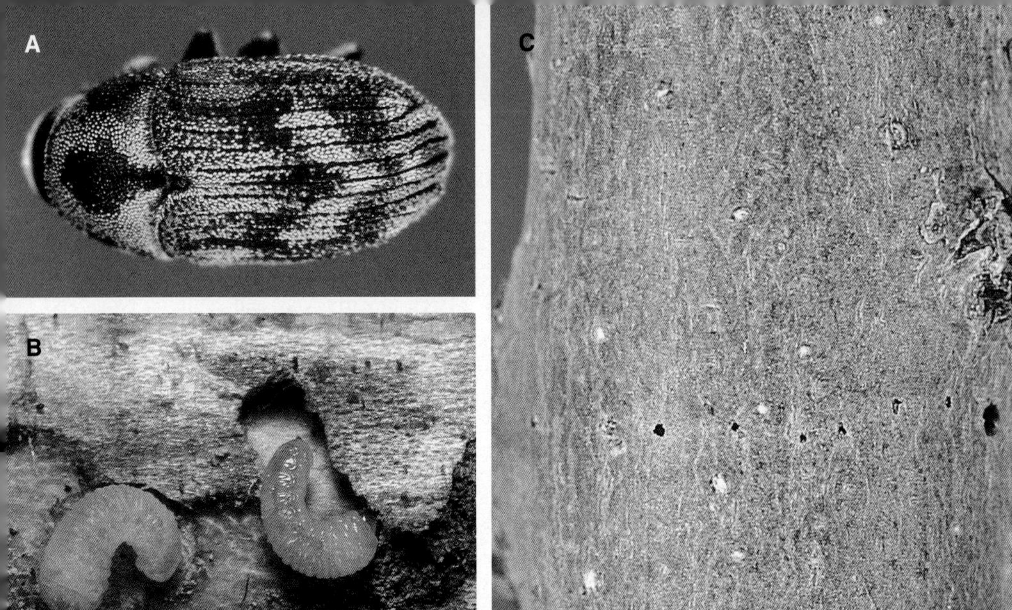
A
B
C

D

E

Symptoms and Signs

Attack by the western ash bark beetle, ***Hylesinus californicus*** (Swaine) (Coleoptera: Scolytidae), can be identified in mid- to late May by the presence of holes, about 2 mm in diameter, in the bark (especially in cracks and crevices), by sap flow from these holes, and by accumulations of boring dust under the attack sites. As adults construct horizontal egg galleries in the phloem, the bark over the galleries turns reddish brown (C) and often becomes sunken and cracked. Beetles occasionally chew ventilation holes, 1–3 mm in diameter and 8–10 mm apart, through the bark over the egg galleries (C). Egg galleries, about 3 mm in diameter and 3–8 cm long, and larval feeding galleries are visible when the covering bark is removed (E). Foliage on twigs and branches girdled by egg galleries turns yellow in late June and early July (D). Adults remain in egg galleries for much of the summer. They are stout-bodied, oval beetles that are 2.0–3.5 mm long; flattened scales covering the body form a dark diamond-shaped patch on the pronotum and a herringbone pattern on the elytra (A). Larvae are creamy-white, legless grubs with light brown heads, and are about 5 mm long when fully grown (B). The eastern ash bark beetle, ***H. aculeatus*** (Say) and Criddle's bark beetle, ***H. criddlei*** (Swaine), look similar to *H. californicus*, and all three species cause similar damage symptoms.

Distribution and Hosts

Hylesinus californicus is distributed from Manitoba to British Columbia and south to northern Mexico; ***H. aculeatus*** and ***H. criddlei*** are distributed from eastern Canada to Alberta and in adjacent states of the United States. All species attack ash (*Fraxinus* spp.). Green ash (*F. pennsylvanica* var. *subintegerrima* (Vahl) Fern.) is most commonly attacked in the prairie provinces.

Life Cycle

Hylesinus californicus has a 1-year life cycle. Adults overwinter in the bark on the lower 15 cm of ash boles. Beetles emerge and disperse to tree crowns in mid-April to May. There they bore through the bark to feed, mate, and reproduce in the phloem of limbs and branches. Each female constructs a transverse egg gallery in the phloem and lays eggs along both sides of the gallery. Larvae hatch in May and feed in the phloem until early July. Larval feeding galleries are perpendicular to the egg gallery (E). Fourth-instar larvae pupate in July and early August. Adults emerge in late July and August, relocate to uninfested parts of branches (usually in a crotch), enter the phloem layer, and feed for several weeks. From mid-September to early November adults stop feeding, emerge from the bark, and migrate by walking, falling, or flying to the base of ash trees, where they burrow into the bark and overwinter. The life cycles of ***H. aculeatus*** and ***H. criddlei*** are similar to that of *H. californicus*.

Damage

Hylesinus californicus is more common and causes more damage than the other two species in the prairie provinces. Damage caused by the three species is usually confined to injured or weakened branches; however, if large amounts of weakened or dead host material are created by wind, snow, ice storms, or drought, beetle populations may reach outbreak proportions. Outbreaks of ***H. californicus*** in southern Alberta and Saskatchewan in the last 10 years are thought to have been precipitated by drought conditions. During outbreaks beetles may also attack and girdle tree trunks, resulting in tree death. Branch mortality also reduces aesthetic value of ornamental and shade trees.

Figure 93. Ash bark beetles. A–E. *Hylesinus californicus*. **A.** Adult. **B.** Larvae. **C.** Green ash branch showing discolored bark and ventilation holes over egg gallery. **D.** Infested green ash tree. Note the yellow foliage on infested branches. **E.** Green ash branch with bark removed to show egg gallery and larval galleries.

A
B
C
D
E
F
G

Symptoms and Signs

Sudden wilting of young leaves in the upper crown is the first visible symptom of Dutch elm disease (DED) in the early summer (C). Wilted leaves shrivel and turn brown but often persist on the tree until winter. When the symptoms appear after the leaves are fully mature, leaves turn yellow and usually drop prematurely. These late-season symptoms are often difficult to distinguish from natural autumn coloring. Dark brown discoloration of the outer sapwood of infected stems is a characteristic internal symptom of the disease (E). Xylem vessels and surrounding cells are discolored and appear as dark rings on cross sections of infected stems. Two kinds of fruiting structures, synnemata and perithecia, can be found in the brood galleries of the beetle. The synnema is a pinhead-like structure having a black stem with white mucilaginous droplets of conidia at the top (F). The perithecium is black with a globose base and a long neck. Ascospores are produced inside perithecia and ooze out from the tip of the neck.

Cause

The pathogen of DED is ***Ophiostoma ulmi*** (Buis.) Nannf. (≡ *Ceratocystis ulmi* (Buis.) C. Moreau, *Ceratostomella ulmi* Buis., anamorph *Pesotum ulmi* (Schw.) Crane & Schoknecht) (Ascomycotina: Ophiostomatales). Recently a new name, ***O. novo-ulmi*** Brasier, has been proposed for aggressive strains in Europe and North America, and nonaggressive strains are kept in *O. ulmi*. Two other wilt diseases of elm (*Ulmus* spp.), caused by ***Dothiorella ulmi*** Verrall & May (Deuteromycotina: Coelomycetes) and ***Verticillium albo-atrum*** Reinke & Berth. (Deuteromycotina: Hyphomycetes), are also associated with the same insects, but they cause much milder disease than *O. ulmi*.

Distribution and Hosts

This disease occurs widely in Europe and North America. It was found for the first time in the prairie provinces in Manitoba in 1975 and has been spreading steadily throughout southeastern Manitoba. In 1981, DED was found in Saskatchewan. The disease is not known in Alberta. In North America DED is spread by two species of bark beetles (Coleoptera: Scolytidae): the native elm bark beetle, ***Hylurgopinus rufipes*** (Eichhoff) (A), and the smaller European elm bark beetle, ***Scolytus multistriatus*** (Marsham) (B). The native elm bark beetle is the major species in the prairie provinces (Manitoba and Saskatchewan). The European elm bark beetle has only been found in a few locations in Manitoba and Saskatchewan; it was found in Alberta (Calgary) for the first time in 1994. The only native elm in the region, American elm (*Ulmus americana* L.), and all introduced elm species are susceptible, but Siberian elm (*U. pumila* L.) and Japanese elm (*U. parvifolia* Jacq.) have some resistance.

Disease Cycle

When bark beetle adults emerge from the bark of diseased trees, they carry spores (mostly conidia) of the pathogen from the galleries to healthy trees. They feed in the crotches of small twigs and deposit spores into the excavated cavities. Once inside the tree, the fungus spreads in the vascular system. Beetles bore beneath the bark of large stems, construct galleries in the cambium layer, and lay eggs (G). Larvae also construct galleries in the bark (F) and produce oval pupal chambers. A toxin called cerato-ulmin produced by the pathogen is responsible for wilting of the tree. Root grafts between diseased and healthy trees also spread DED.

Damage

Trees attacked for several successive years have progressive branch mortality (D). Elm trees of all ages are killed by DED, but overmature, stressed, or weakened trees attract beetles more than young, vigorously growing ones. Mortality of mature trees is a major concern because they are valuable and highly regarded in urban situations, farm shelterbelts, and natural stands.

Figure 94. Dutch elm disease, caused by *Ophiostoma ulmi*. A, B. Two vectors: a native elm bark beetle, *Hylurgopinus rufipes* (A), and a smaller European elm bark beetle, *Scolytus multistriatus* (B). **C.** Recently attacked tree with yellow wilting leaves. **D.** A tree attacked for several years, showing progressive branch death. **E.** Cross section of an infected young stem showing brown discoloration of the outer sapwood. **F.** New larval galleries. **G.** A female gallery with eggs (shiny dots) deposited in the wall.

A
B
C

DIPLODIA GALL AND ROUGH-BARK OF POPLAR

Symptoms and Signs

Globose branch galls of various sizes and the rough-bark symptom occur on trees of all ages (A–C). The bark of infected galls develops deep cracks, which result in the rough-bark appearance (B, C). Small black pycnidial fruiting bodies are often produced on the cracks. On large stems, the infected area consists of raised patches with the rough-bark symptom (B) but no globose galls.

Cause

The cause of Diplodia gall and rough-bark of poplar (*Populus* spp.) is a fungus, ***Diplodia tumefaciens*** (Shear) Zalasky (≡ *Macrophoma tumefaciens* Shear) (Deuteromycotina: Coelomycetes). Another fungus, ***Rhytidiella moriformis*** Zalasky (Ascomycotina: Dothideales), causes the rough-bark symptom on balsam poplar (*P. balsamifera* L.).

Distribution and Hosts

This disease is widely distributed throughout North America, and the major hosts in the prairie provinces are aspen (*Populus tremuloides* Michx.) and balsam poplar. This disease is also known on black cottonwood (*P. trichocarpa* Torr. & Gray), Lombardy poplar (*P. nigra* L. var. *italica* Muenchh.), Brooks poplar (*P. deltoides* Bartr. × Russian), and other poplar hybrids and cultivars.

Disease Cycle

Most of the infections are caused by conidia. An infection begins as a wartlike swelling on a young branch; the swelling grows as the stem increases in size and it becomes a globose gall. On the other hand, an infection on a large branch or stem results only in the rough-bark appearance.

Damage

This disease does not kill trees, but heavy infection does make ornamental trees unsightly and thus reduces their value.

Figure 95. Diplodia gall and rough-bark of poplar, caused by *Diplodia tumefaciens*. A. An aspen heavily attacked by the disease. **B.** Rough-bark symptom on main stem of aspen. **C.** Close-up of young galls.

A
B

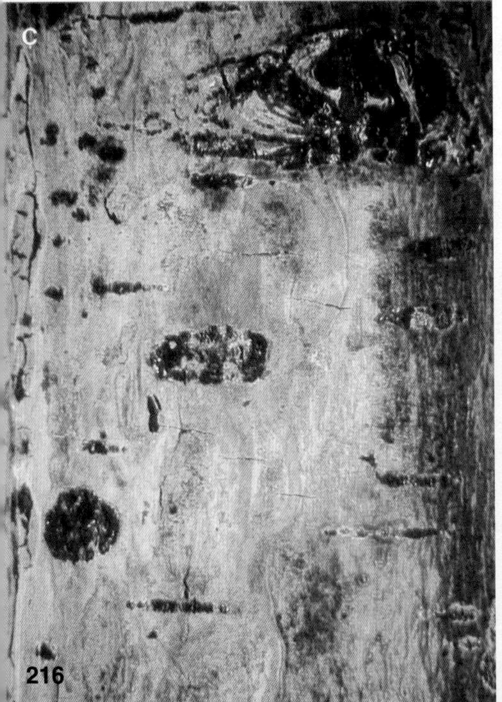
C

D

Symptoms and Signs

The hypoxylon canker starts as a slightly sunken, yellowish orange area on the stem (A). The cankers enlarge rapidly and eventually girdle the stem (B). A mottled or laminated black and yellowish pattern of the cortex and a mycelial fan on the cambium layer are reliable field symptoms and signs of the disease. Older cankers are characterized by black hyphal pegs (C), which are pillarlike structures that push the outer periderm from the underlying cortical tissue. Hyphal pegs are imperfect fruiting structures that produce conidia. After a few years of infection, cankers produce fruiting bodies of the perfect state, which are made up of 20 or fewer perithecia embedded in a round stroma, 5–15 mm in diameter (D). The surface of the stroma is distinctly whitish at first but becomes black with age.

Cause

Hypoxylon canker of aspen is caused by a fungus, ***Entoleuca mammata*** (Wahlenb.) J.D. Rogers & Y.M. Ju (= *Hypoxylon mammatum* (Wahlenb.) P. Karst.) (Ascomycotina: Xylariales).

Distribution and Hosts

Hypoxylon canker is common throughout much of the range of trembling aspen (*Populus tremuloides* Michx.) in North America. The disease also occurs in largetooth aspen (*P. grandidentata* Michx.), balsam poplar (*P. balsamifera* L.), and a few other species of *Populus*.

Disease Cycle

Dissemination of the pathogen takes place by ascospores, and they initiate infection. The role of conidia has not been established, but they may function as spermatia. How the infection starts on aspen is not clearly understood, but the fungus most likely gains entry through injuries on the stem or through dead branch stubs. Lack of moisture and other stresses are likely the important predisposing factors. Insect and woodpecker transmission have been suggested but are not yet confirmed. The pathogen can live as a saprophyte on dead wood and causes white rot; ascospores can be produced for several years after the death of the host trees.

Damage

Hypoxylon canker is considered to be one of the most important aspen diseases in eastern North America, but it is less important in the prairie provinces. This disease is more common in poorly stocked aspen stands, trees under stress, or trees injured by hail, animal, or other means, indicating the secondary nature of the disease. Trees with infection on the lower main stem usually die within 5 years. Hypoxylon canker weakens the stem, which is often broken by wind at the point where the canker occurs.

Figure 96. Hypoxylon canker of trembling aspen, caused by *Entoleuca mammata*. A. A typical elongated canker showing yellow-orange bark discoloration. **B.** A broken aspen tree with Hypoxylon canker. **C.** Close-up of orange-yellow bark symptom. Black hyphal pegs, which are the conidial state of the fungus, are produced beneath the bark. **D.** Several round stromatic fruiting structures made up of 20 or fewer perithecia.

A
B
C
D

E
F

Symptoms and Signs

Symptoms of fire blight include blight of blossom, twig, leaf, collar, and root, and cankers of branch and trunk (A, B). Milky viscous masses of bacteria (B) often ooze out from the cankers, which appear dark and water-soaked. Blossom blight is usually the first symptom of the disease. Blossoms first appear water-soaked and turn brownish to black as they wilt and shrivel (C). Dead and shrivelled blossom clusters and infected fruitlet clusters usually remain attached to the tree and provide good diagnostic clues (D). Twig and leaf blight occurs on young succulent twigs and blackens and wilts them, often forming shepherd's crooks at their tips (D). When infection occurs on immature or mature fruits, the bacteria spread rapidly and rot the fruits. In susceptible hosts, the disease can advance into larger branches or even into main stems (E, F).

Cause

Fire blight is caused by a bacterium, ***Erwinia amylovora*** (Burr.) Winsl. et al. Another bacterial blight caused by ***Pseudomonas syringae*** Van Hall, which causes much less severe damage, also exists in the prairie provinces.

Distribution and Hosts

Fire blight is a native North American disease that occurs throughout the continent, but it has also been recognized in many locations in Europe in recent years. Hosts of fire blight are plants belonging to the family Rosaceae (rose family). Common hosts of the disease in the prairie provinces are apple (*Malus pumila* Mill.), crab apple (*Malus* spp.), pear (*Pyrus* spp.), mountain ash (*Sorbus* spp.), cotoneaster (*Cotoneaster* spp.), saskatoon (*Amelanchier alnifolia* Nutt.), hawthorn (*Crataegus* spp.), and raspberry (*Rubus* spp.).

Disease Cycle

Fire blight bacteria overwinter in infected stem cankers. Bacteria also seem to be able to survive as surface organisms (epiphytes) and can contribute to initial infection in the spring. The bacteria are believed to be carried by wind, rain, and insects from winter cankers to blossoms or young tender shoots in the spring. Blossom infection then moves to the flower stem, the peduncle, and then into the leaves and branches. Secondary infections can occur when bacteria are disseminated by rain, wind, insects, and birds. Also, the disease is often disseminated by human activities through pruning equipment, transported fruits, or nursery stock.

Damage

This is probably the most important disease of ornamental rosaceous trees and shrubs in the prairie provinces. A severe infection can kill mature trees in one season, but some trees can survive for many years after infection.

Figure 97. Fire blight, caused by *Erwinia amylovora*. A. A small canker on crab apple, probably caused by infection through a flower bud. **B.** A canker with a droplet of fire blight bacteria (arrow). **C.** A newly infected crab apple blossom bud showing wilting and discoloration symptoms. **D.** An infected young branch of crab apple showing the shepherd's crook symptom and discolored dead leaves. **E, F.** Heavily infected crab apple.

A
B
220

Symptoms and Signs

Conspicuous greenish brown to black spindle-shaped swellings are the major symptom of black knot of cherry (A, B). These swellings or knots are usually confined to one side of the stem. The surface of the knot is covered by fungal stroma, in which the conidial state and perfect state develop. In the spring the knot is olive-green and the surface has a velvety texture (B), but it turns black and smooth in the fall. Stem swellings are caused by hyperplasia (increased number) of xylem and phloem cells in the infected stems.

Cause

Black knot of cherry (*Prunus* spp.) is caused by a fungus, ***Apiosporina morbosa*** (Schw.) Arx (≡ *Dibotryon morbosum* (Schw.) Theissen & Sydow) (Ascomycotina: Dothideales). A closely related fungus, ***A. collinsii*** (Schw.) Hoehnel, causes black leaf and witches' broom of saskatoon (*Amelanchier alnifolia* Nutt.) (*see* Black Leaf and Witches' Broom of Saskatoon).

Distribution and Hosts

This disease is widely distributed in North America. In Canada, it is known from coast to coast. In the prairie provinces, choke cherry (*Prunus virginiana* L.) and pin cherry (*P. pensylvanica* L.f.) are the main hosts of the disease, but it is also known on many other species of *Prunus* in other parts of North America.

Disease Cycle

During April and May, ascospores, produced in black, cushion-shaped overwintering ascostromata usually, infect the youngest branches and occasionally the large older branches or trunks. Swellings increase in size each year, and spores are produced on their surface every year. The role of conidia is not clear, but they may be involved in dissemination and infection.

Damage

Infected branches are deformed and their growth is significantly reduced. Heavily infected trees often become stunted and deformed.

Figure 98. Black knot of cherry, caused by *Apiosporina morbosa*. A. A heavily infected choke cherry. **B.** Close-up of an infected branch showing a black, spindle-shaped swelling or 'knot'.

A
B
222

Symptoms and Signs

Nectria and ***Cytospora*** are found on dead branches or dead parts of stems. ***Nectria cinnabarina*** (Tode:Fr.) Fr. can be recognized easily by the bright orange to pink raised pustules (sporodochia) (A). ***Cytospora chrysosperma*** Pers.:Fr. can be recognized by the black, multichambered pycnidia (B) producing small, sausage-shaped hyaline spores. Threadlike masses of spores (tendrils) are often observed.

Cause

Many textbooks and publications list fungi such as ***Nectria*** spp. and ***Cytospora*** spp. as canker-causing or dieback pathogens; however, they are likely colonizers of recently killed or weakened stem tissues rather than the primary causes of stem mortality. Common species of such fungi found in the prairie provinces are ***Nectria cinnabarina*** (anamorph *Tubercularia vulgaris* Tode:Fr.) (Ascomycotina: Hypocreales) and ***Cytospora chrysosperma*** (holomorph *Valsa sordida* Nits.) (Deuteromycotina: Coelomycetes). ***Nectria galligena*** Bres. is known to cause target canker on various deciduous tree species in eastern North America but does not occur in the prairie provinces.

Distribution and Hosts

Nectria cinnabarina and ***C. chrysosperma*** are found across the prairie provinces on all common broadleaf tree species.

Disease Cycle

Infections are initiated by both conidia and ascospores, but the details of the disease cycles of the two species are unknown.

Damage

In certain situations, the two fungi are very common and appear to be causing significant damage. Because their pathogenicity is doubtful, they should be considered secondary colonizers of dead or dying trees.

Figure 99. Nectria and Cytospora associated with canker of broadleaf trees. A. *Nectria cinnabarina* on a cherry (*Prunus* sp.). **B.** *Cytospora chrysosperma* on aspen (*Populus tremuloides* Michx.).

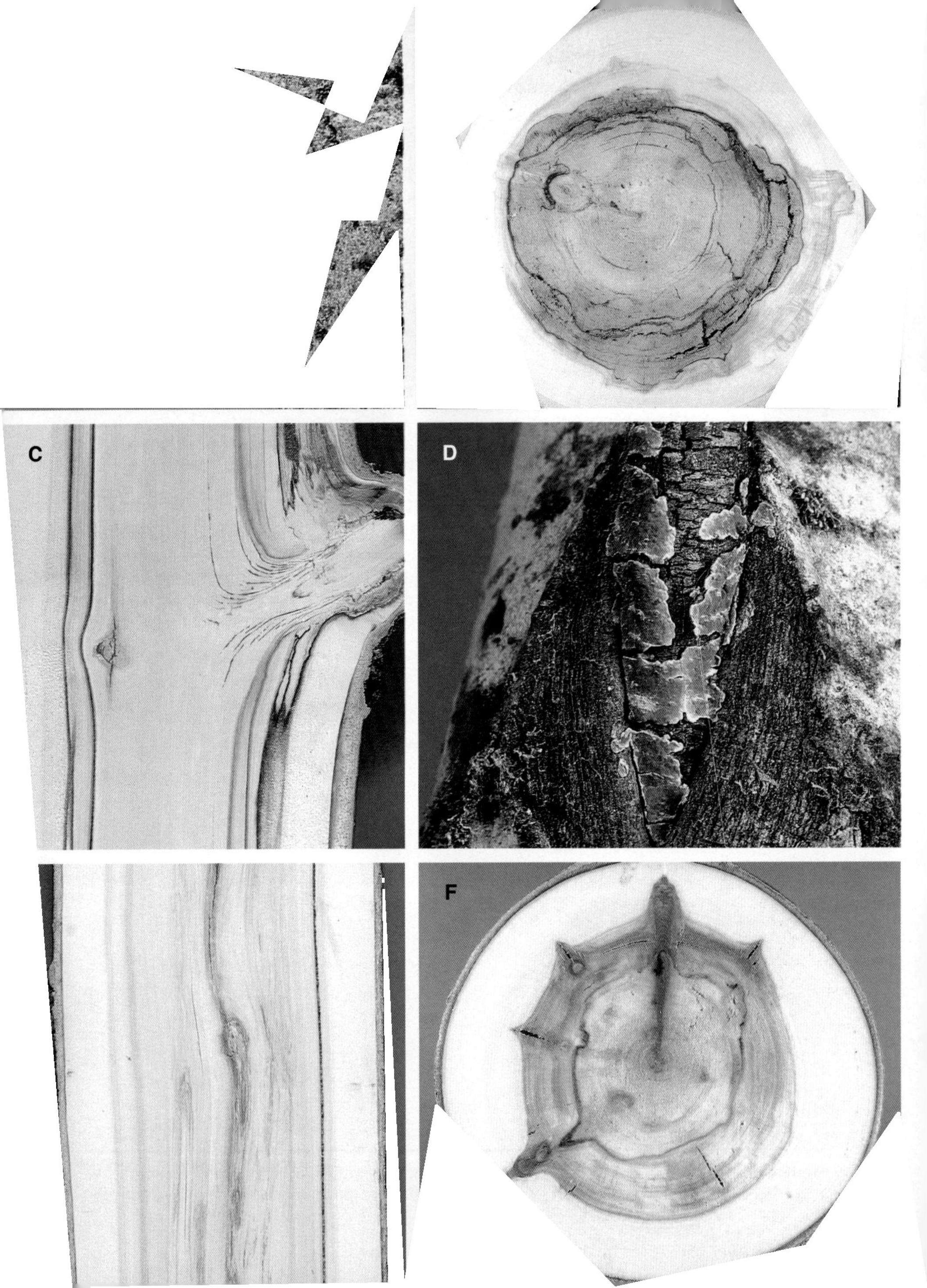
C
D
F

Symptoms and Signs

Preferred part of the wood (heartwood or sapwood), location of decay (e.g., root, butt, trunk), type of decay (white rot or brown rot), or pattern of decay (e.g., pocket rot, stringy rot, or cubical rot) are specific to each decay organism (Table 9). Also, each fungus produces characteristic fruiting bodies (e.g., conks, mushrooms). When fruiting bodies are not associated with the decay, cultural and microscopical characteristics of the fungus are necessary for positive identification.

Cause

Many species of fungi belonging to the subdivision Basidiomycotina are responsible for decay of roots, butts, and trunks of aspen (*Populus tremuloides* Michx.), balsam poplar (*P. balsamifera* L.), and other broadleaf trees in the prairie provinces. More than 250 species of fungi are known to be associated with the decay of aspen in North America. Seventeen of the most common decay fungi of broadleaf trees in the region are listed in Table 9. ***Phellinus tremulae*** (Bond.) Bond. & Boriss. (100A, B, C) and ***Peniophora polygonia*** (Pers.:Fr.) Bourd. & Galzin (100D, E, F) (Aphyllophorales) are the two most economically important decay organisms in standing aspen trees in the prairie provinces.

Distribution and Hosts

All species listed in Table 9 occur throughout the region and are commonly associated with decayed and discolored wood of broadleaf trees.

Disease Cycle

Airborne basidiospores produced on the fruiting bodies (e.g., conks (101A–D) or mushrooms) of decay fungi are the major means of dissemination. New infections are initiated through branch stubs or stem wounds caused by animals, machines, hail, or frost. A few species that cause butt rot and root rot also spread through soil or root contacts. Many decays start only after trees are cut and stored.

Damage

Excessive advanced decay is a factor limiting the economical utilization of broadleaf trees, especially aspen and balsam poplar, in the prairie provinces. Trees with significant advanced decay are not only unmerchantable for structural timber, but because their stems are weak and easily broken by winds or heavy snow, are undesirable as windbreak or amenity trees. Staining of wood or the presence of incipient (early stages of) decay also make the tree undesirable, even though it may be mechanically sound. For pulping, stained wood needs more bleaching chemicals to attain the desired brightness, especially in mechanical processes. Although the amount of decay is known to increase with stand age, other factors, especially extensive stem damage (animal, weather, or mechanical), may also play an important role in the total amount of decay in a stand.

Figure 100. Decay of aspen. A, B, C. *Phellinus tremulae* on aspen. **D, E, F.** *Peniophora polygonia* on aspen.

Figure 101. Decay of balsam poplar and other broadleaf trees. A. *Ganoderma applanatum* on balsam poplar. **B.** *Perenniporia fraxinophila* (Peck) Ryv. on ash (*Fraxinus* spp.). **C.** *Piptoporus betulinus* on birch. **D.** *Fomes fomentarius* on birch (*Betula* sp.).

Table 9. Descriptions of common decay fungi of broadleaf trees in the prairie provinces

Organism	Type of decay	Fruiting body
Armillaria ostoyae (Romagn.) Herink = *A. obscura* (Pers.) Herink **Shoestring root rot, honey mushroom** (*see* Armillaria Root Rot)	Light yellow to white spongy root and butt rot; surface of decay hollow; often with dark brown wall	Honey- to brown-colored gilled mushroom with ring (annulus) on stem (Fig. 52D)
Bjerkandera adusta (Willd.:Fr.) Karst. ≡ *Polyporus adustus* Willd.:Fr. **Scorched conk**	White cubical sap rot	Shelving to sometimes resupinate; up to 7 cm wide; thin upper surface whitish to tan; often dark margins
Cerrena unicolor (Fr.) Murr. ≡ *Daedalea unicolor* Bull.:Fr.	White spongy mottled rot of dead sapwood	Shelving; up to 7 cm wide; upper surface gray, hairy, often with green algae; lower surface whitish with mazelike pores
Coriolus hirsutus (Wulf.:Fr.) Quél. ≡ *Polyporus hirsutus* Wulf.:Fr. **Hairy conk**	White soft spongy rot	Shelving, thin, leathery; up to 5 cm wide; upper surface hairy, gray to brownish gray; lower surface yellowish with small pores
Coriolus versicolor (L.:Fr.) Quél. ≡ *Polyporus versicolor* L.:Fr.	White soft spongy rot of dead sapwood	Shelving, thin, leathery, tough; up to 5 cm wide; upper surface with distinct multicolored (brown, purple, gray) bands with velvety hairs; undersurface white with small pores
Flammulina velutipes (W. Curt.) Karst. ≡ *Collybia velutipes* (W. Curt.) Fr.	White heartrot	Mushroom; up to 7 cm wide; pale buff to orange-brown; centrally stiped
Fomes fomentarius (L.:Fr.) Kickx *Polyporus fomentarius* L.:Fr. **Tinder conk**	Mottled white trunk rot	Perennial, hoof-shaped, tough, woody conk; upper surface light gray; similar to but smaller than *Phellinus tremulae* and paler (Fig. 101D)
Ganoderma applanatum (Pers.) Pat. ≡ *Elfvingia applanatum* (Pers. ex Wallr.) Karst. **Artist's conk**	White mottled spongy root and butt rot	Hard conk up to 70 cm wide; upper surface crusty gray to light brown, dark brown where bruised (Fig. 101A)
Gymnopilus spectabilis (Fr.) A.H. Smith ≡ *Pholiota spectabilis* (Fr.) Kummer	White rot	Yellowish orange gilled mushroom; up to 18 cm in diameter
Hirschioporus pargamenus (Klotzch) Bond. & Singer ≡ *Polyporus pargamenus* Fr.	White pocket sapwood decay, honeycomb appearance on cross sections	Annual, small, thin, shelving; less than 3 cm wide; upper surface white to buff; lower surface deep purple with small tubes

Table 9. Concluded

Organism	Type of decay	Fruiting body
Lyophyllum ulmarium (Bull.:Fr.) Kuehn. ≡ *Pleurotus ulmarius* (Bull.:Fr.) Kummer	White rot	Olive-gray to gray-brown mushroom; up to 25 cm in diameter; eccentric hairy stem; white gills
Peniophora polygonia (Pers.:Fr.) Bourd. & Galzin ≡ *Corticium polygonium* (Pers.:Fr.) Fr.	White rot (Fig. 100E, F)	Pinkish red, disklike patches with smooth hymenial layer; margin white (Fig. 100D)
Phellinus tremulae (Bond.) Bond. & Boriss. = *Fomes igniarius* (L.:Fr.) Fr. **False tinder conk**	White trunk rot with conspicuous black zone lines surrounding each decay column (Fig. 100B, C)	Perennial, hoof-shaped conk; up to 12 cm wide, upper surface blackish, cracked (Fig. 100A)
Pholiota destruens (Brond.) Gill.	White rot	Whitish mushroom; 8–20 cm in diameter; coarse, white woolly scales, domed
Pholiota squarrosa (Pers.) Kummer	White mottled root and butt rot	Small, pale straw-yellow mushroom with rust-brown scales; cap up to 15 cm in diameter
Piptoporus betulinus (Bull.:Fr.) Karst. ≡ *Polyporus betulinus* Bull.:Fr.	Light brown crumbly rot	Shelving; thick, soft, corky conk with lateral stalk; incurved margin; upper surface white to gray (Fig. 101C)
Radulodon americanus Ryv. = *Radulum casearium* (Morg.) Lloyd	White stringy heartrot	Resupinate; teethlike projections; thick (up to 3 cm)

INSECTS AND DISEASES OF BROADLEAF TREES

Seeds

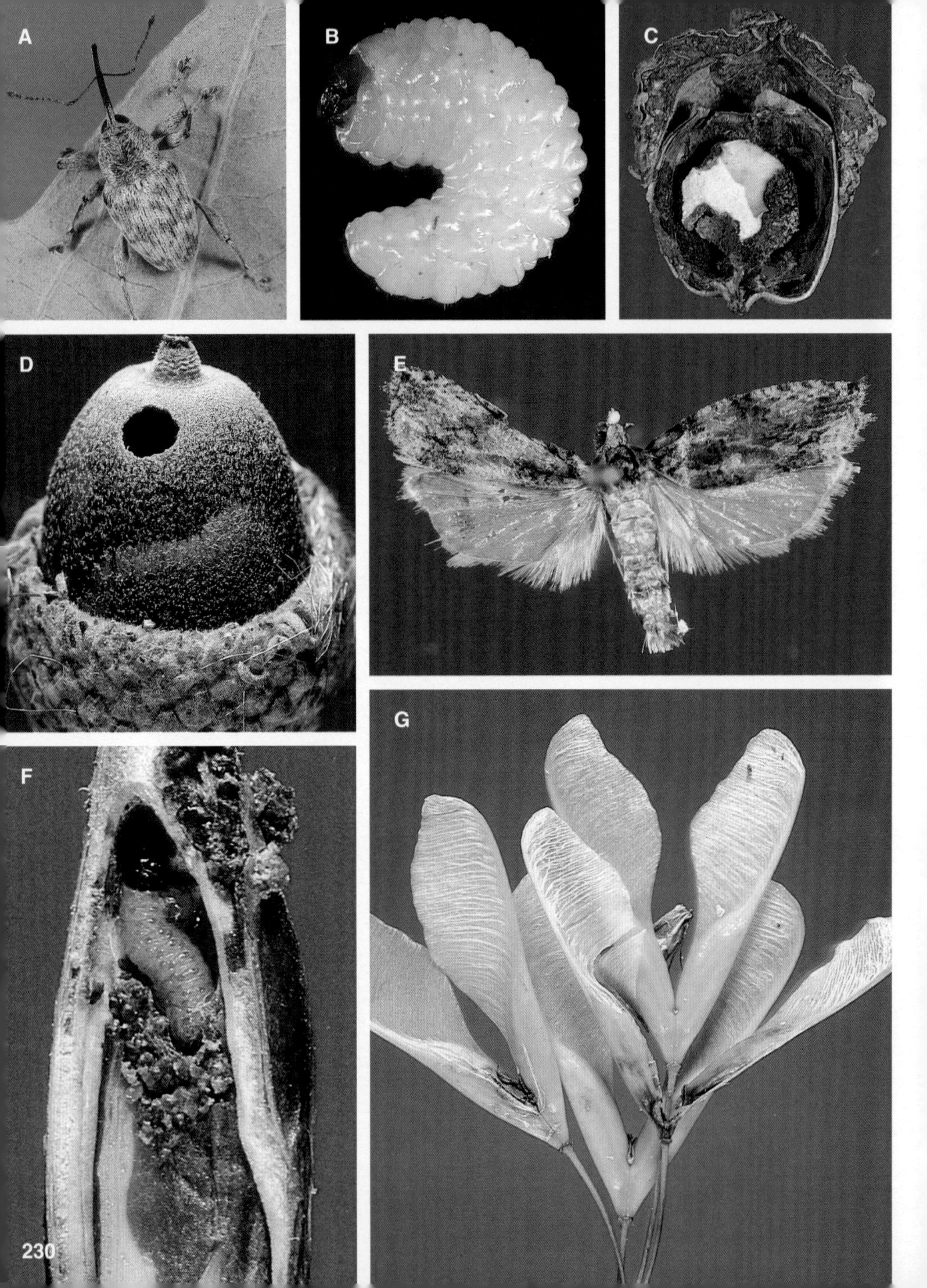
A
B
C
D
E
F
G

Symptoms and Signs

Little is known about insects attacking seeds of hardwood trees in the prairie provinces; however, some species are more common than others. Larvae of a weevil species, ***Curculio iowensis*** (Casey) (Coleoptera: Curculionidae), feed inside acorns. There are no obvious external signs that acorns are infested. When larvae emerge from ripe acorns in late summer they chew round emergence holes, 2–3 mm in diameter, in the top of acorns (D). Larvae are legless, have brown heads and creamy-white, C-shaped bodies, and are about 8 mm long when mature (B). Larvae of a moth species, ***Proteoteras*** nr. ***aesculana*** Riley (Lepidoptera: Tortricidae), burrow into the seeds of Manitoba maple. Burrowing causes an external brownish discoloration of the seeds (G), and larvae expel some frass to the surface of the seeds (F). Larvae and frass are observed when infested seeds are cut open. Larvae are brownish gray caterpillars with blackish heads and thoracic shields, thoracic legs, and five pairs of rudimentary abdominal prolegs, and they are about 10 mm long when fully grown (F).

Distribution and Hosts

Curculio iowensis is found from eastern Canada to Saskatchewan and in adjacent parts of the United States, and attacks oaks (*Quercus* spp.), particularly bur oak (*Q. macrocarpa* Michx.). The distribution of ***P.*** nr. ***aesculana*** is not known, but this species is likely found throughout the natural range of its host, Manitoba maple (*Acer negundo* L.).

Life Cycle

The life cycle of ***C. iowensis*** is poorly known. Adults are present during the summer and lay their eggs in developing acorns. The adults are stout-bodied weevils with long snouts and mottled light and dark gray bodies, and they are 8–10 mm long (A). Mature larvae chew emergence holes in the acorns (D), emerge, and drop to the ground to pupate.

Proteoteras nr. ***aesculana*** has a 1-year life cycle. Adults are small moths with wingspans of about 12 mm. The front wings are mottled brown or gray and the hind wings are pale (E). Moths are active from mid-May to late June. Females lay eggs near the veins of leaves or on the base of young seeds. Larvae hatching from the latter burrow directly into seeds, eventually destroying the contents. Larvae emerge from seeds in early August and drop to the ground to overwinter. It is not known if pupae or larvae overwinter.

Damage

Curculio iowensis larvae eat all or part of acorn contents (C), destroying the viability of seeds. Populations of this species are occasionally high in Manitoba and may destroy 10–20% of acorns in a year. ***Proteoteras*** nr. ***aesculana*** also kills seeds and may affect a high proportion of seeds on some trees.

Figure 102. **Seed insects. A–D.** *Curculio iowensis*. **A.** Adult. **B.** Larva. **C.** Bur oak acorn damaged by larval feeding. **D.** Larval emergence hole in bur oak acorn. **E–G.** *Proteoteras* nr. *aesculana*. **E.** Adult. **F.** Manitoba maple seed opened to show young larva and damage. **G.** Manitoba maple seeds infested with larvae.

OTHER DAMAGING AGENTS

Weather

A

B

C

D

Symptoms

Symptoms of late-spring frost damage are wilting, discoloration, and dropping of current year's foliage (A, B, C). Broadleaf trees such as aspen (*Populus tremuloides* Michx.) flush again after the spring frost damage but produce foliage with a patchy appearance (D) because they do not reflush uniformly. Sometimes damage to new shoots of spruce by late-spring frost (B) is confused with shoot-tip swelling caused by Cooley spruce gall adelgid (***Adelges cooleyi*** (Gillette)); however, shoots infested with the adelgid are swollen (*see* Adelgid Galls on Spruce), whereas shoots killed by late-spring frost are not. Shoot damage to spruce (*Picea* spp.) and pine (*Pinus* spp.) by terminal weevils (***Pissodes*** spp.) (*see* White Pine Weevil, and Lodgepole Terminal Weevil) can also be confused with late-spring frost damage.

Cause

Late-spring frost damage occurs in the spring when the temperature drops below freezing after new foliage has already developed but is still young and succulent. When frost occurs in late spring or early summer, usually only young, succulent foliage is damaged and the mature, hardened foliage remains unharmed (B).

Damage

The damage can be extensive in young exposed stands, but trees usually produce new shoots and recover. Repeated spring frost damage in successive years often results in deformed trees (A) and substantial growth loss.

Figure 103. **Late-spring frost damage. A.** Mortality of new growth at various heights in small white spruce (*Picea glauca* (Moench) Voss). **B.** Typical drooping symptoms on white spruce. **C.** Previous year's needles of lodgepole pine (*Pinus contorta* Dougl. var. *latifolia* Engelm.) killed by late-spring frost. **D.** Clumping of aspen leaves caused by late-spring frost damage followed by second flushing.

A

B

Symptoms

The symptoms of winter desiccation damage are discoloration of foliage or parts of trees, and dieback (A, B). Winter desiccation damage can be confused with damage caused by Armillaria root rot or root collar weevil because of similar reddish discoloration. The basal portion of trees affected by winter desiccation usually remains green, whereas trees affected by Armillaria root rot or root collar weevil are completely discolored, and there is no snow-line delineation (*see* Armillaria Root Rot, and Root Collar Weevils).

Cause

Death of whole trees or various degrees of dieback damage occur during the winter on coniferous trees as a result of winter desiccation. The damage is also known as frost drought. This kind of damage usually occurs when tree parts that are aboveground or above the snow line are exposed to warm dry air and sun, which cause excessive transpiration, while the ground is still frozen. Often the snow-cover height at the time of damage is visible on affected trees (A, B). Needles under the snow cover are protected from damage because of a lower rate of transpiration. Red-belt damage, which will be discussed separately, is another example of winter drying damage. Two types of winter desiccation damage can be distinguished: chronic and acute. Chronic winter desiccation develops slowly and is due to gradual water loss by transpiration during the 3- to 4-month period in winter when no water is available from the roots because of soil frost. Acute winter desiccation damage arises from a breakdown of the water balance, and symptoms can appear within a few days. This happens when unusually warm and dry weather conditions occur when the ground is still frozen.

Damage

Damage caused by winter desiccation can be extensive and significant. Economic loss is great when high-value seedlings for horticultural use or Christmas trees are affected. The damage is more common on exotic than native tree species. However, buds are usually not killed and trees recover within a few years.

Figure 104. Winter desiccation damage. A. Lodgepole pine (*Pinus contorta* Dougl. var. *latifolia* Engelm.) showing snow-line delineation. **B.** White spruce (*Picea glauca* (Moench) Voss) showing snow-line delineation.

A
B

Symptoms

Symptoms of red belt usually appear in early spring on forested slopes as areas of reddish brown discoloration of various shapes and sizes (A, B). South- and west-facing slopes are most affected by red belt. On long slopes the damage often appears as horizontal, well-defined bands, thus the name 'red belt'. More commonly, damage occurs in distinct patches in the middle of slopes along mountain ridges, and occasionally in valley bottoms. Red belt also appears along the fringes of stand openings.

Needles and buds of a whole tree can be killed, but buds are usually not killed and the affected tree recovers within a few years. Often only a part or one side of a tree is affected.

Cause

The precise cause of red belt is not known. It is thought to result from a combination of climatological conditions; no infectious biological agent is involved. Drastic change of air temperature during the winter caused by warm chinook conditions is thought to be responsible for red belt.

Distribution and Hosts

Red belt damage occurs mainly in the upper foothills of the eastern slopes of the Rocky Mountains in the prairie provinces, but it is also known to occur in British Columbia and the western United States. Although all native conifers are affected, lodgepole pine (*Pinus contorta* Dougl. var. *latifolia* Engelm.) is the most susceptible tree species in this region.

Damage

Although mortality is rare, very severe or repeated red belt damage at the same location could kill significant numbers of trees. Mortality of approximately 7000 ha of mature and immature lodgepole pine in the Cadomin area of the Alberta upper foothills of the Rocky Mountains was attributed to red belt in 1971. Trees weakened by red belt become vulnerable to bark beetle attack or root rot infection. Red belt has little economic significance because it usually occurs on steep slopes in inaccessible areas. When trees are killed over large areas, however, dead needles make them extremely flammable and increase fire hazard.

Figure 105. **Red belt. A.** An extensive red belt affecting most of a hillside. **B.** Two bands of red belt on a hill.

A

B

C

D

Symptoms

Typical symptoms of needle droop are sharp downward bending and dying of current season's needles. Drooping occurs when the needles are still green (A, B). Later, the needles die and gradually turn reddish brown (C, D); most of them remain attached to the stem well into the next summer. Depending on when the stress occurs and the maturity of the needles, either the top parts of the trees (B, D) or the lower parts of the trees (A, C) may be affected.

Cause

The phenomenon of needle droop typically affects red pine (*Pinus resinosa* Ait.) in Minnesota, Wisconsin, Ontario, and Manitoba, especially planted stock. Needle droop is caused by abnormal water flow to trees. If sudden and excessive rapid transpiration occurs when soil has limited moisture supply, an absorption lag follows. Succulent tissue in the needle base collapses and causes the needles to droop.

Damage

Significant numbers of planted red pine seedlings can be killed or deformed by needle droop, depending on the climate and soil conditions.

Figure 106. **Needle droop of red pine. A.** Needle droop on the bottom half of planted red pine. **B.** Typical needle droop symptoms. **C.** Discolored needles several months after the damage. **D.** Close-up of discolored needles.

A

B

Symptoms

Bent or broken branches, toppling of trees, shredding or dropping of foliage, and stem scars are symptoms of weather-related damage to trees. This kind of damage is sometimes difficult to diagnose because the cause is not present at the time of diagnosis. Often trees having structurally weakened stems due to stem diseases (e.g., pine stem rusts, canker diseases, extensive heartrots) and root and butt rots (e.g., Armillaria root rot, Tomentosus root rot) are severely damaged.

Cause

Severe wind storms including tornados, heavy snow, and ice glaze can cause mechanical damage to trees.

Figure 107. Wind and snow damage. A. Snow damage to white spruce (*Picea glauca* (Moench) Voss). **B.** Wind damage to jack pine (*Pinus banksiana* Lamb.) enhanced by Tomentosus root rot.

A
B
C
D

Symptoms

Physical damage such as broken branches, shredded foliage, and open stem wounds of various sizes are evident soon after hailstorms. Damage to stems from previous storms is usually characterized by open blisters surrounded by healing ribs. Old hail damage can be confused with stem cankers or blisters caused by fungi or insects, but it is possible to determine the original cause of damage. Hail damage always occurs on the side of the tree from which the wind was blowing, while damage from other causes usually occurs on all sides of the stem.

Cause and Damage

Hailstorms can occur throughout the year but are less frequent in the winter. The size of the area in which the damage occurs depends on the extent and severity of the hailstorm.

Damage

Young stands can be completely demolished by a severe hailstorm but impact to larger trees is often insignificant. Hail scars eventually heal, but in the interim they may provide entrance courts for pathogenic fungi.

Figure 108. **Hail damage. A.** A stand of lodgepole pine (*Pinus contorta* Dougl. var. *latifolia* Engelm.) severely damaged by hail. **B.** Hail damage on the main stem of a lodgepole pine. **C.** Severe hail damage of young balsam poplar (*Populus balsamifera* L.). **D.** Typical healing of hail scars on a lodgepole pine branch.

A
B

Symptoms

Large areas of forests can be affected by drought, and in severe cases extensive mortality occurs. Red pine (*Pinus resinosa* Ait.) and jack pine (*Pinus banksiana* Lamb.) needles turn reddish brown (A, B). Drought damage is often difficult to diagnose because the symptoms are similar to those caused by many diseases and insects such as Armillaria root rot, Scleroderris canker, and root collar weevil. Drought-stressed trees are predisposed to attack by many diseases and insects. For several years after severe drought damage, various degrees of dieback and tree mortality continue even though the drought conditions have disappeared.

Cause and Damage

If very dry conditions prevail for several growing seasons, trees in a wide area may suddenly show foliage discoloration symptoms. Striking mortality of red pines in many plantations in the Belair and Sandilands areas of Manitoba from 1985 to 1989 is believed to have been caused by drought conditions that existed several years before the damage. Drought damage and winter desiccation should be differentiated. Drought damage is due to a lack of adequate moisture in the soil, whereas winter desiccation occurs when moisture in the soil is temporarily unavailable because of frozen ground, even if moisture is present in the soil.

Figure 109. **Drought damage. A.** Extensive drought damage of planted red pines. **B.** Drought damage of a natural jack pine stand.

A
C

LIGHTNING DAMAGE—CIRCULAR TREE MORTALITY

Symptoms

Although many lightning strikes occur without igniting forest fires, lightning is the most prevalent cause of forest fires in the prairie provinces. Groups of dead trees in circular patches (A, B, C) have been seen from time to time, especially in the foothills of the Rocky Mountains. This phenomenon is called 'circular tree mortality' or 'group tree mortality'.

Tree mortality occurs in a single circle or several circles side by side. The circles are usually between 0.1 and 0.2 ha, and about 40 to 150 trees are discolored and killed in each circle. One can usually locate a tree in the center of the circle with evidence of the lightning strike such as a burnt stem or a burnt black streak along one side of the tree trunk. Circular tree mortality can be mistaken for stand-opening root rot diseases such as Armillaria root rot or Tomentosus root rot, but trees in circular tree mortality all die in one season rather than gradually from the center over the years, as occurs with root rot diseases. Circular tree mortality of pines has also been mistaken for mountain pine beetle damage.

Cause and Damage

The cause of circular tree mortality has not been proven in the prairie provinces, but is likely caused by lightning strikes, as observed and documented in the United States and Australia. Circular mortality of a similar nature has also been reported in wheat, potato, and peanut fields. The damage caused by lightning strikes is also known to attract bark beetles, which infest surrounding trees and cause more damage.

Figure 110. **Lightning damage—circular tree mortality. A.** A distant view of circular tree mortality in a lodgepole pine forest (*Pinus contorta* Dougl. var. *latifolia* Engelm.). **B.** Affected lodgepole pine within a circle. **C.** Close-up view of circular tree mortality within a lodgepole pine stand.

OTHER DAMAGING AGENTS

Chemicals

A

B

C

D

DAMAGE CAUSED BY HERBICIDES, SOIL STERILANTS, AND OTHER AGRICULTURAL CHEMICALS

Symptoms

Symptoms vary depending on the kind and concentration of the chemicals, frequency of use, species and condition of trees, and the timing of chemical application. Some of the common symptoms are wavy or curled leaf margins; chlorosis; abnormal swelling, twisting, and curling of branchlets (A); blight and necrosis of leaves (B); and browning and casting of conifer needles (C, D) or deciduous tree leaves.

Cause

Damage caused by misuse or overuse of herbicides (A, B), soil sterilants (C, D), and other agricultural chemicals has increased conspicuously during recent years. Many herbicides are based on plant hormones that interrupt or unbalance normal growth. Under certain conditions, nontarget plants, such as planted trees, may be affected by herbicide application, especially if application prerequisite conditions are not met.

Distribution and Hosts

All tree species are susceptible to damage by herbicides and other agricultural chemicals. Damage is common on urban trees, farm shelterbelts, and forest trees close to agricultural fields and under power lines where herbicides are applied frequently. In recent years, herbicide use has been considered in managed forest stands and plantations in Canada, but there are concerns about hazards, efficacy, and cost.

Damage

Improper use of agricultural chemicals may cause partial or total mortality of trees. Indirect effects to mycorrhizal fungi and to shrub tree species such as buffalo-berry (*Shepherdia* spp.), sweet gale (*Myrica gale* L.), and alder (*Alnus* spp.), which are known to aid forest soil fertility with actinorrhizae, should also be considered.

Figure 111. **Damage caused by herbicides and soil sterilants. A.** Herbicide damage to willow (*Salix* spp.). **B.** Foliar symptoms of aspen (*Populus tremuloides* Michx.), caused by a herbicide spray. **C.** Several white spruce (*Picea glauca* (Moench) Voss) affected by soil sterilant treatment along a fence line. **D.** Domestic white spruce affected by soil sterilant application.

A
B
C
D
E
F
G
254

Symptoms

The type of damage and severity of symptoms caused by chemical pollutants depend on the kind, concentration, and exposure time to the pollutants and the species, age, health, vigor, and seasonal developmental stage of host plants. Some of the major symptoms associated with chemical pollutant damage are chlorosis (C), necrosis, discoloration, wilting, dieback, and premature defoliation (G). A slow progressive decline in the health of trees over large areas has been reported in the eastern United States and Europe; chemical pollution is thought to be involved. However, there is no widespread or regional forest decline in the prairie provinces that can be attributed to pollution.

Cause

When trees are exposed for a long time to high concentrations of chemical pollutants in the environment they may die or exhibit various types of damage. The pollutants may be gaseous, liquid, or solid, being emitted from one or many industrial sources, including oil and gas extraction facilities, coal mining, mineral mining and smelting, and the pulp and paper industry (A–G). Exhaust gas from gasoline and diesel engines, mainly from automobiles, also contains a significant amount of chemical pollutants. Another source of chemical pollution results when large amounts of gaseous oxides of nitrogen and sulfur are dissolved in rainwater, producing acid rain or acid precipitation. The following are the main chemical pollutants that can harm trees, along with their major sources:

Natural gas and oil refining: sulfur dioxide, hydrogen sulfide, liquid hydrocarbons, saltwater spills

Metal mining and smelting: sulfur dioxide, heavy metals, particulates, fluorides

Other minerals: cement, potash

Pulp and paper industry: hydrogen sulfide, methyl mercaptan, dimethyl sulfide, sodium hydroxide, ammonia

Stationary combustion engines: nitrogen oxides, ozone, PAN (peroxyacetyl nitrate)

Distribution and Hosts

Damage to forest trees by chemical pollutants is usually restricted to areas close to industries emitting toxic substances or along heavily traveled highways; however, the effects of acid rain are known to spread over long distances and wide areas. In the prairie provinces most natural gas wells are located on the eastern slopes of the Rockies; oil sand plants are in northern Alberta; oil refineries are near Edmonton, Regina, and Calgary; pulp and paper companies are located in Grande Prairie, Hinton, Whitecourt, Peace River, and Athabasca in Alberta, Meadow Lake and Prince Albert in Saskatchewan, and The Pas and Pine Falls in Manitoba; potash mines are located in southern Saskatchewan; and heavy metal processing plants are located in the Flin Flon and Thompson areas of Manitoba.

Damage

Damage to forest trees caused by chemical pollutants does not appear to be significant in the prairie provinces at the present time except in localized areas; however, careful monitoring of areas around major point source locations is important to forewarn of future damage.

Figure 112. **Chemical pollutant damage. A.** Acute SO_2 fumigation injury to lodgepole pine (*Pinus contorta* Dougl. var. *latifolia* Engelm.). **B.** Close-up of acute SO_2 fumigation injury on lodgepole pine, showing symptom development from the needle tips. **C.** Typical fluoride injury to jack pine (*P. banksiana* Lamb.) near an aluminum smelter. **D.** Elemental sulfur damage to ground-cover vegetation near sulfur piles. **E.** Extensive tree mortality around a well blowout. **F.** Symptoms on aspen (*Populus tremuloides* Michx.) from a single acute exposure to metal smelter emissions. **G.** Jack pine affected by emissions from a pulp mill.

OTHER DAMAGING AGENTS

Birds and Animals

A
B
C
D
E
F

Symptoms

Typical symptoms of damage by various animals are given below:

Root clipping and basal stem debarking: vole
Stem debarking of young trees: snowshoe hare, squirrel
Seedling clipping: vole, snowshoe hare, squirrel
Foliage clipping and browsing: elk, deer, moose
Mature tree debarking: bear, elk, deer, porcupine (A, B, F)
Large tree cutting: beaver (C)
Seed destruction: white-footed deer mouse, meadow vole, chipmunk, shrew
Cone (branch) clipping: squirrel (D, E)

Cause

Many species of small and large mammals damage forest trees in the prairie provinces (A–F). The two animals known to cause the most damage to young conifer trees in managed forests are snowshoe hare (***Lepus americanus*** Erxleben) and red squirrel (***Tamiasciurus hudsonicus*** Erxleben). Other animals that damage forest trees in this region are porcupine (***Erethizon dorsatum*** Cuvier), beaver (***Castor canadensis*** Khul.), elk (***Cervus canadensis nelsoni*** Bailey), deer (***Odocoileus*** spp.), moose (***Alces alces*** Gray), and bear (***Euarctos*** spp. and ***Ursus*** spp.). Small mammals such as white-footed deer mouse (***Peromyscus maniculatus borealis*** Mearns.), meadow vole (***Microtus pennsylvanicus*** Ord), red-backed vole (***Clethrionomys gapperi*** Vigors), chipmunk (***Eutamias*** spp.), and shrew (***Sorex*** spp.) are known to damage conifer seeds on the ground.

Distribution and Hosts

Intensity, distribution, and types of animal damage are dependent on the population level and kinds of animals involved, availability of other food sources for the animals, species and age of trees, and climatic and biotic conditions of the general area. Some species of animals have definite annual patterns of population increase and decrease. For example, the snowshoe hare is known to reach peak population levels every 9–10 years.

Damage

Damage by animals, especially by snowshoe hare and red squirrel, is significant in young managed conifer stands. The two animals may cause cumulative damage of 40–50% of young trees. In addition to mortality and disfigurement of trees, animal feeding leaves scars that often provide entry courts for canker-causing fungi or wood decay fungi. Stem rust cankers of pines are selectively gnawed by rodents and rabbits. Significant amounts of seeds on the forest floor can be destroyed by small animals before germination. This is considered to be the most important problem facing direct seeding operations for conifers.

Figure 113. **Animal damage. A.** Extensive bark injury of lodgepole pine (*Pinus contorta* Dougl. var. *latifolia* Engelm.) caused by porcupine. **B.** Bark injury to lodgepole pine, probably caused by deer. **C.** A balsam poplar (*Populus balsamifera* L.) cut down by beaver. **D.** Close-up of injury caused by conc clipping by squirrel (arrow). **E.** Red flagging of lodgepole pine caused by cone clipping by squirrel. **F.** Debarking injury of lodgepole pine caused by bear.

A
B
C

Symptoms

Sapsuckers drill rows of holes on the bark of young trees through to the sapwood (A–C) and visit these trees periodically to lap up the sap that has oozed out of the holes. Holes drilled by sapsuckers are always arranged in regular patterns.

Cause

The yellow-bellied sapsucker (***Sphyrapicus varius varius*** (L.)) is a migratory member of the woodpecker family measuring 20–25 cm long. The crown and throat of the bird are crimson; the rest of the head is black with white streaks above and below the eye extending onto the neck; the lower breast and abdomen are yellowish. A related species, the red-bellied sapsucker (***S. varius ruber*** (Gmelin)), is found in coastal British Columbia.

Distribution and Hosts

This bird lives throughout the mixed-wood forests of North America and is often seen during the summer. The most common tree species in the prairie provinces, birch (*Betula* spp.), aspen (*Populus tremuloides* Michx.), and pines (*Pinus* spp.), can be severely damaged by the yellow-bellied sapsucker.

Damage

If many holes are drilled by the bird, the trees can be weakened. Because sapsucker feeding-trees are often found near the margins of recently logged areas or exposed sites near roadsides, there is some indication that the birds prefer weakened trees. Sapsucker wounds provide entry points for stain-causing and decay fungi.

Figure 114. Sapsucker injury. A, B. Extensively injured jack pine (*Pinus banksiana* Lamb.). **C.** Extensively injured birch.

REFERENCES

Borror, D.J.; Triplehorn, C.A.; Johnson, N.F. 1989. An introduction to the study of insects. 6th ed. Saunders College Publ., Philadelphia, Pennsylvania.

Dix, M.E.; Pasek, J.E.; Harrell, M.O.; Baxendale, F.P., tech. coords. 1991. Common insect pests of trees in the Great Plains. U.S. Dep. Agric., For. Serv., and Univ. Nebraska Coop. Ext. Serv., Lincoln, Nebraska. Great Plains Agric. Counc. Publ. 119.

Evans, D. 1982. Pine shoot insects common in British Columbia. Pac. For. Res. Cent., Victoria, British Columbia. Inf. Rep. BC-X-233.

Farr, D.F.; Bills, G.F.; Chamuris, G.P.; Rossman, A.Y. 1989. Fungi on plants and plant products in the United States. Am. Phytopathol. Soc., St. Paul, Minnesota.

Finck, K.E.; Humphreys, P.; Hawkins, G.V. 1990. Field guide to pests of managed forests in British Columbia. For. Can., Pac. Yukon Reg., Pac. For. Cent., Victoria, British Columbia, and B.C. Minist. For., Protect. Branch, Victoria, British Columbia. Joint Publ. 16.

Funk, A. 1981. Parasitic microfungi of western trees. Can. For. Serv., Pac. For. Res. Cent., Victoria, British Columbia. Inf. Rep. BC-X-222.

Funk, A. 1985. Foliar fungi of western trees. Can. For. Serv., Pac. For. Res. Cent., Victoria, British Columbia. Inf. Rep. BC-X-265.

Furniss, R.L.; Carolin, V.M. 1977. Western forest insects. U.S. Dep. Agric., For. Serv., Washington, D.C. Misc. Publ. 1339.

Gerber, H.S.; Tonks, N.V.; Ross, D.A. 1980. The recognition and life history of the major insect and mite pests of ornamental shrubs and shade trees of British Columbia. B.C. Minist. Agric., Victoria, British Columbia. Revised.

Ginns, J.H. 1986. Compendium of plant disease and decay fungi in Canada 1960-1980. Agric. Can., Res. Branch, Ottawa, Ontario. Publ. No. 1813.

Hagle, S.K.; Tunnock, S.; Gibson, K.E.; Gilligan, C.J. 1987. Field guide to diseases and insect pests of Idaho and Montana forests. U.S. Dep. Agric., For. Serv., State and Private For., North. Reg., Missoula, Montana.

Hedlin, A.F. 1974. Cone and seed insects of British Columbia. Environ. Can., Can. For. Serv., Pac. For. Res. Cent., Victoria, British Columbia. Inf. Rep. BC-X-90.

Hiratsuka, Y. 1987. Forest tree diseases of the prairie provinces. Can. For. Serv., North. For. Cent., Edmonton, Alberta. Inf. Rep. NOR-X-286.

Hiratsuka, Y.; Powell, J.M. 1975. Pine stem rusts of Canada. Environ. Can., Can. For. Serv., Ottawa, Ontario. Publ. 1329.

Hiratsuka, Y.; Zalasky, H. 1993. Frost and other climate-related damage of forest trees in the prairie provinces. For. Can., Northwest Reg., North. For. Cent., Edmonton, Alberta. Inf. Rep. NOR-X-331.

Holsten, E.H.; Hennon, P.E.; Werner, R.A. 1985. Insects and diseases of Alaskan forests. U.S. Dep. Agric., For. Serv., For. Pest Manage., State and Private For., Alaska Reg., Juneau, Alaska. Rep. 181. Revised.

Hosie, R.C. 1979. Native trees of Canada. 8th ed. Fitzhenry & Whiteside Ltd., Don Mills, Ontario, and Can. For. Serv., Ottawa, Ontario.

Ives, W.G.H.; Wong, H.R. 1988. Tree and shrub insects of the prairie provinces. Can. For. Serv., North. For. Cent., Edmonton, Alberta. Inf. Rep. NOR-X-292.

Johnson, W.T.; Lyon, H.H. 1991. Insects that feed on trees and shrubs. 2nd ed. Revised. Comstock Publishing Associates, Ithaca, New York.

Malhotra, S.S.; Blauel, R.A. 1980. Diagnosis of air pollutant and natural stress symptoms on forest vegetation in western Canada. Environ. Can., Can. For. Serv., North. For. Res. Cent., Edmonton, Alberta. Inf. Rep. NOR-X-228.

Martineau, R. 1984. Insects harmful to forest trees. Multiscience Publications Ltd., Montreal, Quebec.

Moss, E.H. 1959. Flora of Alberta. Univ. Toronto Press, Toronto, Ontario.

Myren, D.T., ed. 1994. Tree diseases of eastern Canada. Nat. Resour. Can., Can. For. Serv., Sci. Sustainable Devel. Directorate, Ottawa, Ontario.

Riffle, J.W.; Peterson, G.W., tech. coord. 1986. Diseases of trees in the Great Plains. U.S. Dep. Agric., For. Serv., Rocky Mtn. For. Range Exp. Stn., Fort Collins, Colorado. Gen. Tech. Rep. RM-129.

Sinclair, W.A.; Lyon, H.H.; Johnson, W.T. 1987. Diseases of trees and shrubs. Comstock Publishing Associates, Ithaca, New York.

Stein, J.D.; Kennedy, P.C. 1972. Key to shelterbelt insects in the northern Great Plains. U.S. Dep. Agric., For. Serv., Rocky Mtn. For. Range Exp. Stn., Fort Collins, Colorado. Res. Pap. RM-85.

Stoetzel, M.B. 1989. Common names of insects and related organisms. Entomological Society of America, Lanham, Maryland. Rev. ed.

Sutherland, J.R.; Miller, T.; Quinard, R.S., eds. 1987. Cone and seed diseases of North American conifers. North American Forestry Commission, Victoria, British Columbia. Publ. 1.

Sutherland, J.R.; van Eerden, E. 1980. Diseases and insect pests in British Columbia forest nurseries. B.C. Minist. For., Victoria, British Columbia, and Environ. Can., Can. For. Serv., Pac. For. Res. Cent., Victoria, British Columbia. Joint Rep. 12.

Wong, H.R.; Melvin, J.C.E.; Harper, A.M. 1977. Common insect and mite galls of the Canadian prairies. Fish. Environ. Can., Can. For. Serv., North. For. Res. Cent., Edmonton, Alberta. Inf. Rep. NOR-X-196.

Ziller, W.G. 1974. The tree rusts of western Canada. Dep. Environ., Can. For. Serv., Ottawa, Ontario. Publ. 1329.

The capital letters in some definitions refer to Figure 115 (pp. 270–271).

Abdomen: The part of an insect's body located posterior to the thorax (C, D, E, G, J, K, L).

Acervulus (pl. acervuli): A mat of hyphae bearing a cushion of short conidiophores and conidia.

Actinorrhiza (pl. actinorrhizae): The symbiotic association of Actinomycetes with the roots of higher green plants in which the organism lives in close association with root cells, especially those at the root surface, and often produces root nodules.

Aeciospore: A spore with two nuclei that is produced in an aecium of a rust fungus.

Aecium (pl. aecia): A sorus of a rust fungus that produces aeciospores; often associated with spermogonia.

Alternate host: One or the other of the two different hosts of a host-alternating (heteroecious) rust fungus or insect (some aphids and adelgids).

Amphigenous: The occurrence of the fruiting bodies of a parasitic fungus on both surfaces of the leaves of an infected plant.

Anal shield: A hard plate on the dorsal posterior segment of caterpillars and certain other immature insects (F, H).

Anamorph: The nonsexual, or imperfect, state of a fungus; compare holomorph and teleomorph.

Annual: Lasting only one year or one growing season.

Annulus (annular ring): A ringlike veil on the stem of a mushroom.

Apothecium (pl. apothecia): A cup- or disc-shaped ascocarp.

Appressed: Pressed closely to or lying flat against something.

Ascocarp: The ascus-bearing organ of an ascomycete.

Ascomycete: A fungus belonging to the subdivision Ascomycotina.

Ascospore: The sexually produced spore of an ascomycete.

Ascostroma (pl. ascostromata): A type of ascocarp, which is a mass of fungal tissue containing asci.

Ascus (pl. asci): A saclike cell of an ascomycete within which ascospores are produced.

Autoecious: A fungus completing its life cycle on one kind of host; compare heteroecious.

Bacterium (pl. bacteria): A unicellular microscopic organism that lacks chlorophyll and multiplies by fission.

Basal cups: Cuplike structures of dwarf mistletoe that remain on the branches of the host tree after the aerial shoots fall off.

Basidiomycete: A fungus belonging to the subdivision Basidiomycotina.

Basidiospore: A spore produced on a basidium.

Basidium (pl. basidia): In a basidiomycete, a spore-bearing structure in which meiosis occurs.

Biclavate: Clublike on both ends, with a constricted center section.

Bifusiform: Both ends spindle shaped, with a constricted center section; dumbbell shaped.

Bifusoid: Both ends somewhat spindle-shaped with constricted center section.

Blight: Sudden and severe damage to leaves, flowers, and stems.

Blotch mine: Irregularly circular- or oval-shaped damage on a leaf caused by a leaf-mining insect feeding on the mesophyll but leaving the epidermis.

Burl: A hemispherical growth on a tree stem.

Cambium: A meristematic layer of cells lying between xylem and phloem.

Canker: A necrotic, often sunken lesion on a stem, branch, or twig of a plant.

Cellulose: A complex carbohydrate and fundamental constituent of cell walls of all green plants.

Cerato-ulmin: A toxin produced by the pathogen of Dutch elm disease and thought to be responsible for wilting of infected trees.

Chitin: The principal chemical component of an insect's exoskeleton (outer covering).

Chlorosis: Yellowing of normally green tissue due to destruction of chlorophyll or failure of chlorophyll to develop.

Chrysalis (pl. chrysalides): The pupa of a butterfly.

Clavate: Club shaped.

Cleistothecium (pl. cleistothecia): An enclosed ascocarp without an opening; typical of a powdery mildew; compare perithecium.

Cocoon: A protective chamber made from silk and/or host materials constructed by a final-instar larva, and inside which pupation occurs.

Concolorous: Of the same color.

Conidium (pl. conidia): An asexual propagating spore.

Conk: Common term for a fruiting body of a wood-decay fungus.

Context: In a basidiocarp, the hyphal mass between the upper surface and the subhymenium.

Crawler: An actively moving immature stage of an adelgid or scale insect.

Cuticle: A continuous noncellular layer covering the aerial parts of a higher plant.

Damping-off: Death of seeds or seedlings caused by soil-borne fungi.

Decay: The gradual decomposition of dead wood tissue.

Diapause: A part of the life cycle of many insects in which there is minimal metabolic activity and no growth.

Dieback: Death of tree tissue beginning at branch tips and progressing toward the stem, or starting at the top of the stem and progressing downward.

Dioecious: Having male and female organs on different individuals of the same species.

Disease: Any malfunctioning of host cells and tissues that results from continuous irritation by a pathogenic agent or environmental factor and leads to development of symptoms.

Disease cycle: The chain of events involved in disease development, including the stages of development of the pathogen and the effect of the disease on the host.

Distal: Toward the apex or tip of a structure.

Dorsal: Pertaining to the upper surface of an insect or structure (F, H).

Eccentric: Off-center.

Effused-reflexed: Flat basidiocarp on the substratum, with turned-up edges.

Ellipsoid: A mass that is round in cross section and elliptical in side view.

Elliptical: Of elongated circular shape; shape of an ellipse.

Elytral declivity: The downcurved area (I) at the posterior end of the elytra that may be flat, concave, bumpy, or spiny.

Elytrum (pl. elytra): One of the hard or leathery front wings of a beetle (I) that form a protective covering for the abdomen and functional hind wings.

Endophytic: Situated within plant tissues.

Epidermis: The outer cell layer of higher plants.

Epiphyte: An organism living on the surface of another organism, but not as a parasite.

Erumpent: Breaking through the host tissue.

Exoskeleton: The hard outer covering of an insect.

Falcate: Curved like the blade of a scythe or sickle.

Filiform: Threadlike.

Flocculence: A woolly or fuzzy, whitish covering of soft wax produced by some adelgids.

Frass: Wood fragments mixed with excrement produced by a boring insect.

Fruiting body: Fungal structure containing or bearing spores.

Fungus (pl. fungi): An organism without chlorophyll, usually having a mycelium and sexual and/or asexual spores.

Fusiform: Spindle shaped, narrowing toward the ends.

Gall: Outgrowth or swelling that is often more or less spherical; caused by insects, fungi, or other organisms.

Generation: The period of time between the birth of one organism and that of its offspring.

Genital aperture: An opening at the posterior apex of the abdomen of female insects through which eggs or larvae are laid.

Germ pore: An area, or hollow, in a spore wall (especially in rust fungi) through which a germ tube may grow.

Globose: Spherical or almost so.

Hard pines: Pines (*Pinus* spp.) with leaves in clusters of two or three, persistent basal sheaths, and cone scales often with prickles; also called yellow pines or pitch pines.

Heartwood: The central part of a tree that does not conduct water.

Hemicellulose: Group of polysaccharides that accompany cellulose and lignin in wood cell walls.

Hemielytrum (pl. hemielytra): One of the front wings of true bugs (order Hemiptera); they are thickened and leathery at the base, but very thin and membranous at the apex (A).

Herbicide: A chemical that kills weeds or weed seeds.

Heteroecious: A fungus completing its life cycle on two kinds of hosts; compare autoecious.

Hibernaculum (pl. hibernacula): The silken shelter spun by the larvae of some moths, and in which hibernation of the larva occurs.

Holomorph: A fungus having both the sexual and asexual spore states; compare anamorph and teleomorph.

Honeydew: A sugary fluid excreted by feeding aphids, scale insects, and some other insects in the order Homoptera.

Host: Any plant attacked by a pathogen or insect.

Hyaline: Transparent, colorless.

Hygroscopic: Capable of responding to changes in moisture.

Hymenium (pl. hymenia): The spore-producing layer of a sporophore of a basidiomycete.

Hyperplasia: The production of an abnormally large number of cells; compare hypertrophy.

Hypertrophy: An abnormal increase in the size of cells; compare hyperplasia.

Hypha (pl. hyphae): The threadlike filament of a fungus.

Hyphal peg: A peglike bundle of hyphae arising from a cushion of fungus, typical of the conidial state of the Hypoxylon canker pathogen.

Hypodermis: A layer of cells below the epidermis of coniferous needles.

Hysterothecium (pl. hysterothecia): An elongated ascocarp of a needle cast fungus, usually with a central slit.

Imperfect state: The asexual state of a fungus; also called an anamorph.

Instar: The period of development of an insect larva or nymph between molts.

Larva (pl. larvae): The immature stage between the egg and pupal stages of an insect with complete metamorphosis. Caterpillars and maggots are examples of larvae (D, F, H, J, L).

Life cycle: The series of changes in the form of an organism, beginning with the egg and ending with the adult reproductive stage.

Lignin: Complex aromatic substance in woody cell walls and between them that imparts mechanical strength.

Mesophyll: The middle or internal tissues of a leaf, between the epidermal layers.

Midrib: The central vein of a leaf.

Mold: The common name for a visible fungus growing on a surface.

Molt: The process of shedding the exoskeleton by an insect or mite.

Molting tent: A temporary silken shelter spun by some insect larvae to protect them while molting.

Mushroom: A large fruiting body of a fungus, with a fleshy cap and a stalk.

Mycangium (pl. mycangia): A structure (often a pocket or cavity) in the exoskeleton of some bark- and wood-boring insects that contains symbiotic fungi, especially yeasts.

Mycelium (pl. mycelia): A mass of fungal hyphae.

Mycoparasite: A fungus parasitic on another fungus.

Mycorrhiza (pl. mycorrhizae): The symbiosis of certain fungi with the roots of higher green plants in which the fungus lives in close association with the root cells, especially those at the root surface.

Necrosis: Localized death of cells or tissues.

Needle cast: Loss or casting of needles of coniferous trees.

Nymph: The immature stage between the egg and adult stages of an insect with incomplete metamorphosis. Nymphs are usually similar to the adult in general body form, but are smaller and lack fully developed wings and functional genitalia (B).

Ovipositor: The external egg-laying apparatus of some female insects.

Ovoid: Shaped like an egg.

Parasite: An organism living in or on another living organism and obtaining its nutrients from the host; compare saprophyte.

Parthenogenesis: The ability of females to reproduce without fertilization; a form of asexual reproduction.

Pathogen: An organism capable of inciting disease.

Peduncle: The stalk of a flower or flower cluster.

Perennial: Lasting more than one year.

Perfect state: The sexual state of a fungus; also called a teleomorph.

Periderm: The outermost corky layer of the bark of a tree.

Perithecium (pl. perithecia): A flask-shaped ascocarp of an ascomycete, with a conspicuous top opening; compare cleistothecium.

Petiole: The stem that attaches a leaf to a twig or shoot.

Pheromone: A chemical substance released into the air by an insect to communicate with others of the same species.

Phytoplasma: A unicellular microscopic organism without an organized nucleus or a true cell wall; also called mycoplasma-like organism (MLO).

Phloem: Vascular tissue that conducts synthesized foods through plants; occurs outside of the cambium layer in trees; compare xylem.

Powdery mildew: A fungus belonging to the order Erysiphales of the subdivision Ascomycotina.

Prepupa (pl. prepupae): A stage of the final larval instar that is preparing to molt into a pupa. Prepupae usually do not feed and movement is reduced.

Proleg: One of the fleshy legs on the abdomen of most caterpillars and sawfly larvae. The number of prolegs may be used to distinguish sawfly larvae (six or more pairs of prolegs) from caterpillars (five or fewer pairs) (D, F, H).

Pronotum (pl. pronota): The dorsal part of the first thoracic segment, often enlarged or otherwise structurally distinct from the rest of the thorax (A, B, I).

Pseudosclerotium (pl. pseudosclerotia): A compact mass of fungal mycelium intermixed with soil or plant tissue; a survival mechanism of a fungus.

Pseudothecium (pl. pseudothecia): An ascostromatic ascocarp having asci in unwalled cavities (locules).

Pupa (pl. pupae): The developmental stage between a larva and adult in an insect with complete metamorphosis.

Puparium (pl. puparia): A hard protective covering surrounding a fly pupa and produced by the inflation and hardening of its final larval exoskeleton.

Pycnidium (pl. pycnidia): A flask-shaped conidia-producing structure usually with an ostiole (mouth).

Resinosis: Excessive outflow of resin from coniferous plants; usually resulting from injury or disease.

Resupinate: A flat basidiocarp with a spore-producing outer layer.

Rhizomorph: A root- or cord-like bundle of hyphae.

Rot: Disintegration and decomposition of plant tissue accompanied by discoloration.

Rust: A fungus belonging to the order Uredinales of the subdivision Basidiomycotina.

Saprophyte: An organism using dead organic material as food; compare parasite.

Sapwood: The water-conducting outer portion of a tree stem inside the cambium.

Sclerotium (pl. sclerotia): A firm, often dark-colored and rounded sterile mass of fungal hyphae; a survival mechanism of a fungus.

Semiglobose: Semispherical.

Septate: Having cross-walls or partitions.

Septum (pl. septa): A wall, cross-wall, or partition.

Seta (pl. setae): A bristle-like hair on the body surface of an insect.

Shelving: Flat fruiting bodies of fungi that grow one on top of the other, similar to shelves.

Sign: A characteristic structure of the causal agent of a disease; compare symptom.

Skeletonize: A type of defoliation caused by insects in which all leaf tissue is consumed except the leaf veins.

Soft pines: Pines (*Pinus* spp.) with leaves in clusters of five, deciduous basal sheaths, and cone scales mostly without prickles; also called white pines or five-needle pines.

Sordid: Of a dull or muddy color.

Sorus (pl. sori): A structure that produces spore masses; especially in rust fungi.

Spermatium (pl. spermatia): A noninfecting sexual gamete of a fungus.

Spermogonium (pl. spermogonia): A spermatium-producing organ.

Spore: A general term for a reproductive propagule of fungi, bacteria, and lower plants.

Sporodochium (pl. sporodochia): A cushionlike mass of conidiophores.

Sporophore: A spore-producing or spore-supporting structure.

Spumaline: The gray, foamy, hardened protective covering on egg bands of tent caterpillars.

State: A phase of the fungus life cycle.

Stipe: A stem or stalk of a fruiting body of a fungus.

Stroma (pl. stromata): A mass of fungal tissue packed tightly.

Subcuticular: Produced under the cuticle and above the epidermis of a leaf.

Subepidermal: Produced under the epidermis of a leaf.

Subhypodermal: Produced under the hypodermis of coniferous needles.

Symptom: A visible reaction of a host plant as a result of a disease; compare sign.

Synnema (pl. synnemata): A cemented fascicle of erect conidiophores, with a head of conidia-bearing branches; also called a coremium.

Systemic: Distributed throughout an organism.

Teleomorph: The sexual (perfect) state of a fungus; compare anamorph and holomorph.

Teliospore: A spore state of a rust fungus that germinates to produce a basidium.

Telium (pl. telia): A sorus that bears teliospores.

Thoracic shield: A hard plate on the dorsal part of the first thoracic segment (F) of some caterpillars and other larvae. The color of this shield may be different from that of the rest of the abdomen.

Thorax: The middle section of an insect's body, where the wings and legs are attached; in legless, immature insects, the first three segments after the head capsule (C, D, E, J, K, L).

Tubercle: A small knoblike or rounded fleshy protuberance (H) of the exoskeleton.

Urediniospore: A repeating vegetative spore of a rust fungus.

Uredinium (pl. uredinia): A sorus of a rust fungus that produces urediniospores.

Vacuole: A cavity in the cytoplasm, usually filled with fluid.

Vector: An organism (usually an insect) that can transmit a disease.

Ventral: Pertaining to the lower surface of an insect or structure (F).

Verrucose: Having small, rounded processes or warts.

Virus: A submicroscopic parasite usually consisting of nucleic acid surrounded by a protein coat.

Viscin: A clear, sticky, tasteless substance from the mucilaginous sap of the mistletoe.

Wilt: Loss of freshness or drooping of plants; also a disease causing the wilting symptom.

Witches' broom: A proliferation of branches in a dense cluster caused by fungi or other causes.

Xylem: The water-conducting tissue of higher plants; in trees it is produced inward from the cambium; compare phloem.

Zone line: Narrow, dark brown or black lines or plates in decayed wood; generally caused by fungi.

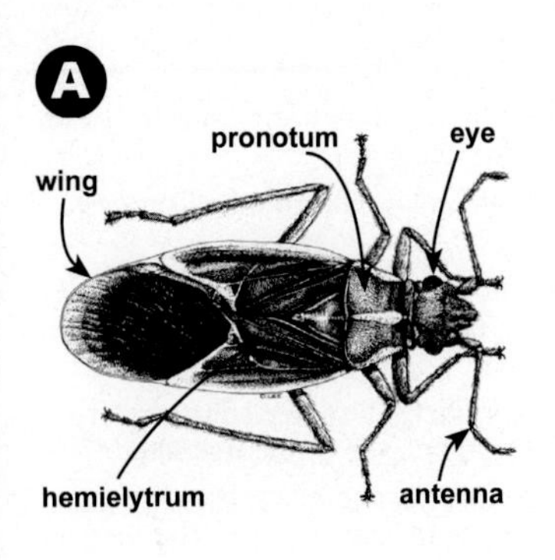
A
wing
pronotum
eye
hemielytrum
antenna

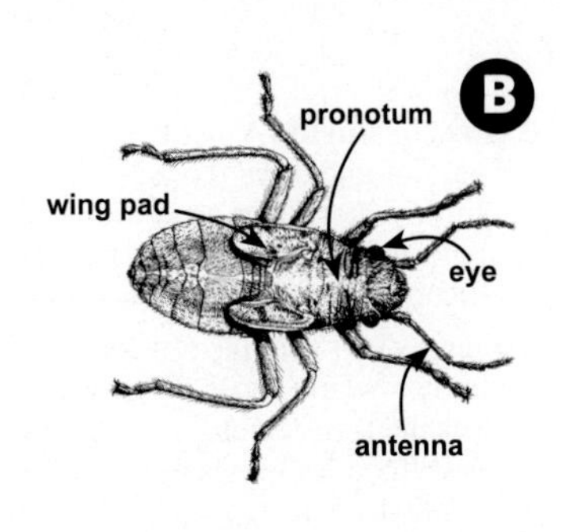
B
pronotum
wing pad
eye
antenna

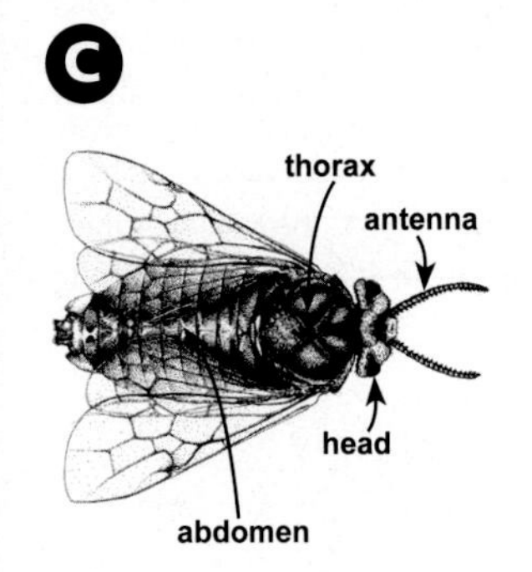
C
thorax
antenna
head
abdomen

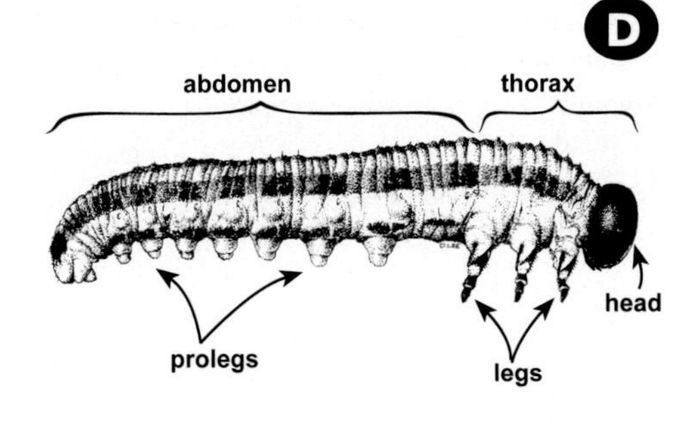
D
abdomen
thorax
head
prolegs
legs

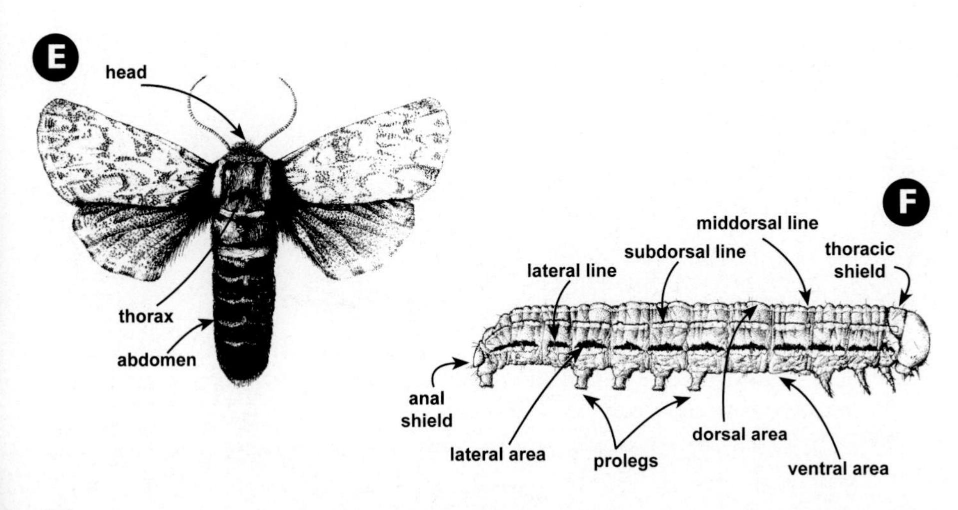
E
head
thorax
abdomen
F
middorsal line
subdorsal line
thoracic shield
lateral line
anal shield
dorsal area
lateral area
prolegs
ventral area

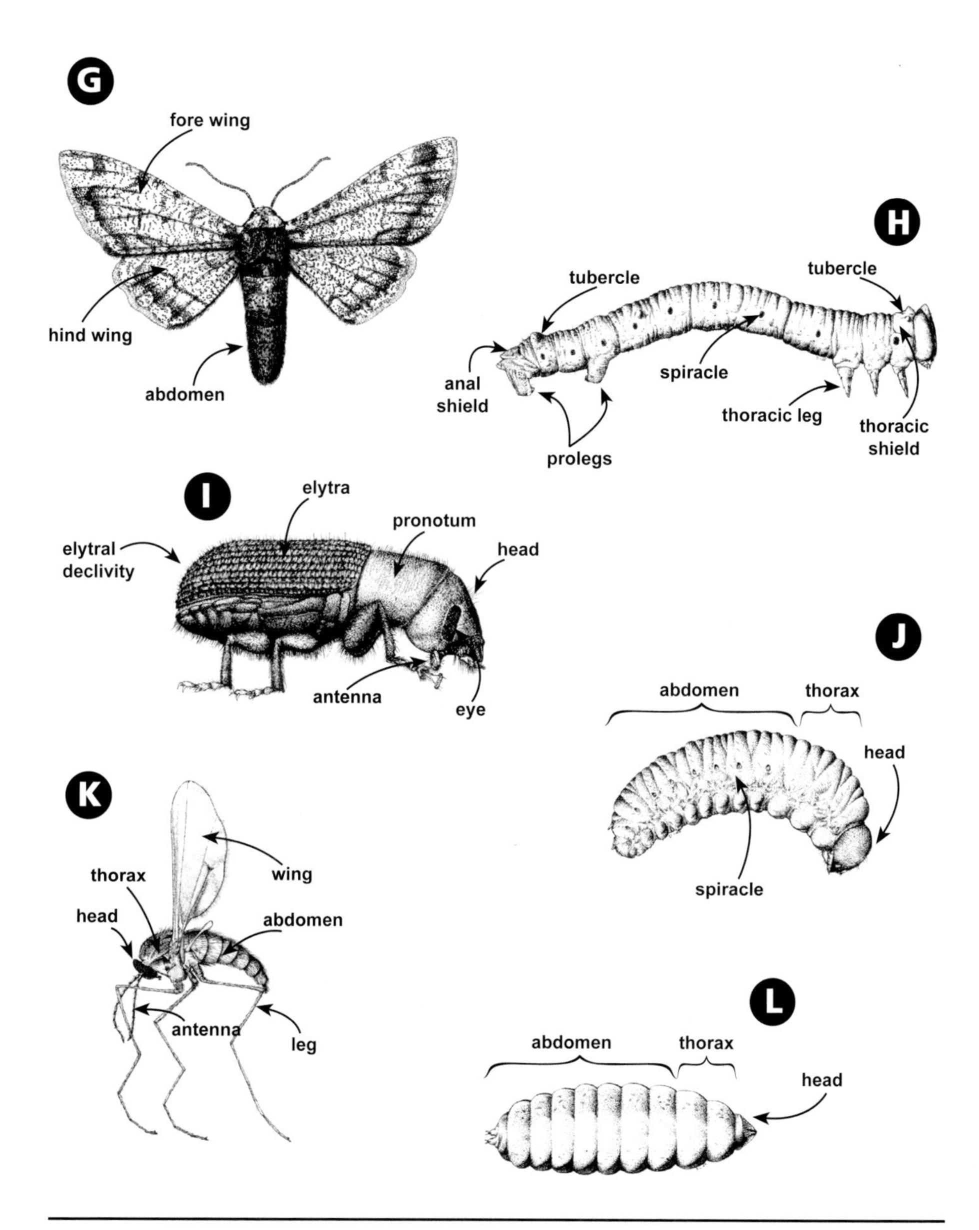

Figure 115. Some terms used to describe insects. A. Plant bug adult (Hemiptera). **B.** Plant bug nymph. **C.** Sawfly adult (Hymenoptera). **D.** Sawfly larva. **E.** Moth (Lepidoptera). **F.** Moth larva or caterpillar. **G.** Looper moth (Lepidoptera). **H.** Looper larva or caterpillar. **I.** Bark beetle adult (Coleoptera). **J.** Bark beetle larva. **K.** Midge adult (Diptera). **L.** Midge larva.

This index lists insects, mites, and diseases (fungi, bacteria, mistletoes) by host and part of host affected. Other biotic and abiotic agents that affect trees are not included in this index, but are described on pages 233-261.

***Abies* spp. (fir)**
***balsamea* (L.) Mill. (balsam fir)**
***lasiocarpa* (Hook.) Nutt. (alpine fir, subalpine fir)**
Foliage and buds
Insects
Choristoneura fumiferana (spruce budworm), 5
Neodiprion abietis (balsam fir sawfly), 15
Zeiraphera canadensis (spruce budmoth), 9
see also Conifers
Diseases
Herpotrichia juniperi (brown felt blight), 31
Hyalopsora aspidiotus (needle rust), 53
Isthmiella abietis (needle cast), 49, 51
Isthmiella quadrispora (needle cast), 49, 51
Lirula abietis-concoloris (needle cast), 49, 51
Lophomerum autumnale (needle cast), 49, 51
Melampsora abieti-capraearum (needle rust), 53
Melampsorella caryophyllacearum (witches' broom rust), 53
Nothophacidium abietinellum (snow blight), 49, 51
Phacidium abietis (snow blight), 49, 51
Phaeocryptopus nudus (needle fungus), 49, 51
Pucciniastrum epilobii (needle rust), 53
Pucciniastrum goeppertianum (needle rust), 53
Sarcotrochila balsameae (snow blight), 49, 51
Uredinopsis phegopteridis (needle rust), 53
Roots, stems, and branches
Insects
Monochamus notatus (northeastern sawyer beetle), 65
Tetropium cinnamopterum (wood borer), 67
see also Conifers
Diseases
see Conifers
Seeds and cones
Insects
Megastigmus specularis (balsam fir seed chalcid), 131
see also Conifers
Seedlings
see Conifers

***Acer* spp. (maple)**
***negundo* L. (Manitoba maple)**
Foliage and buds
Insects
Archips negundana (larger boxelder leafroller), 151, 152, 153
Contarinia negundifolia (boxelder leaf gall midge), 169
Leptocoris trivittatus (boxelder bug), 167
Periphyllus negundinis (boxelder aphid), 167
see also Broadleaf trees
Diseases
Microsphaera penicillata (powdery mildew), 185
see also Broadleaf trees
Roots, stems, and branches
see Broadleaf trees
Seeds
Insects
Proteoteras nr. *aesculana* (caterpillar), 231
Seedlings
see Broadleaf trees

***Alnus* spp. (alder)**
Foliage and buds
Insects
Altica spp. (flea beetles), 159
Phratora spp. (leaf beetles), 159
see also Broadleaf trees
Diseases
Erysiphe aggregata (powdery mildew), 185
Microsphaera penicillata (powdery mildew), 185
Phyllactinia guttata (powdery mildew), 185
see also Broadleaf trees

Roots, stems, and branches
Insects
Cryptorhynchus lapathi (poplar and willow borer), 201
see also Broadleaf trees
Diseases
see Broadleaf trees

Amelanchier alnifolia Nutt. (saskatoon, serviceberry)
Foliage and buds
Insects
Eriosoma americanum (woolly elm aphid), 169
see also Broadleaf trees
Diseases
Apiosporina collinsii (black leaf and witches' broom), 191
Gymnosporangium clavariiforme (leaf and berry rust, juniper stem rust), 187, 189
Gymnosporangium clavipes (leaf and berry rust, juniper stem rust), 187, 189
Gymnosporangium nelsonii (leaf rust, juniper gall rust), 187, 189
Gymnosporangium nidus-avis (leaf rust, causes witches' brooms, juniper witches' broom rust), 187, 189
Phyllactinia guttata (powdery mildew), 185
Roots, stems, and branches
Diseases
Erwinia amylovora (fire blight), 219
see also Broadleaf trees

Arctostaphylos rubra (Rehder & Wils.) Fern. (alpine bearberry)
Pucciniastrum sparsum (spruce–bearberry rust), 43, 45

Arctostaphylos uva-ursi (L.) Spreng. (kinnikinnick, common bearberry)
Chrysomyxa arctostaphyli (leaf rust, yellow witches' broom of spruce), 43, 45, 47

Aster spp. (aster)
Coleosporium asterum (pine needle rust), 39

Betula spp. (birch)
papyrifera Marsh. (white birch)
Foliage and buds
Insects
Altica spp. (flea beetles), 159
Arge pectoralis (birch sawfly), 155
Bucculatrix canadensisella (birch skeletonizer), 161
Enargia decolor (aspen twoleaf tier), 151, 152, 153
Fenusa pusilla (birch leafminer), 163
Heterarthrus nemoratus (late birch leaf edgeminer), 163
Oligocentria lignicolor (lacecapped caterpillar), 151, 152, 153
Phratora spp. (leaf beetles), 159
Profenusa thomsoni (ambermarked birch leafminer), 163
see also Broadleaf trees
Diseases
see Broadleaf trees
Roots, stems, and branches
Insects
Agrilus anxius (bronze birch borer), 203
Cryptorhynchus lapathi (poplar and willow borer), 201
see also Broadleaf trees
Diseases
see Broadleaf trees
Seedlings
see Broadleaf trees

Broadleaf trees (Hardwood trees)
Foliage and buds
Insects
Alsophila pometaria (fall cankerworm), 149
Archips argyrospila (fruit tree leafroller), 151, 152, 153
Archips cerasivorana (uglynest caterpillar), 147
Caliroa cerasi (pear sawfly), 161
Choristoneura conflictana (large aspen tortrix), 139
Cimbex americana (elm sawfly), 155
Corythucha spp. (lace bugs), 167
Datana ministra (yellownecked caterpillar), 151, 152, 153
Eriophyes spp. (gall mites), 171
Hyalophora cecropia (cecropia moth), 151, 152, 153
Hyphantria cunea (fall webworm), 147
Leucoma salicis (satin moth), 145
Lochmaeus manteo (variable oak leaf caterpillar), 151, 152, 153
Lymantria dispar (gypsy moth), 143

Malacosoma disstria (forest tent caterpillar), 137, 147
Nymphalis antiopa (mourningcloak butterfly), 151, 152, 153
Operophtera bruceata (Bruce spanworm), 141
Paleacrita vernata (spring cankerworm), 149
Diseases
Chondrostereum purpureum (silver leaf), 193
Pseudomonas syringae (bacterial blight), 219
Roots, stems, and branches
Insects
Chionaspis furfura (scurfy scale), 209
Quadraspidiotus perniciosus (San José scale), 209
Lepidosaphes ulmi (oystershell scale), 209
Parthenolecanium corni (European fruit lecanium), 209
Diseases
Armillaria calvescens (Armillaria root rot), 115
Armillaria mellea (Armillaria root rot), 115
Armillaria ostoyae (Armillaria root rot), 109, 110, 115, 225, 227
Armillaria sinapina (Armillaria root rot), 115
Bjerkandera adusta (scorched conk), 225, 227
Cerrena unicolor (white spongy mottled rot), 225, 227
Chondrostereum purpureum (silver leaf), 193
Coriolus hirsutus (hairy conk), 225, 227
Coriolus versicolor (white soft spongy rot), 225, 227
Cytospora chrysosperma (Cytospora canker), 223
Flammulina velutipes (white heartrot), 225, 227
Fomes fomentarius (tinder conk), 225, 226, 227
Ganoderma applanatum (artist's conk), 225, 226, 227
Gymnopilus spectabilis (white rot), 225, 227
Hirschioporus pargamenus (white pocket sapwood decay), 225, 227
Lyophyllum ulmarium (white rot), 225, 228
Nectria cinnabarina (Nectria canker), 223
Nectria galligena (target canker), 223
Perenniporia fraxinophila (white rot), 225, 226
Pholiota destruens (white rot), 225, 228
Pholiota squarrosa (white mottled root and butt rot), 225, 228
Piptoporus betulinus (light brown crumbly rot), 225, 226, 228
Pseudomonas syringae (bacterial blight), 219
Radulodon americanus (white stringy heartrot), 225, 228
Seedlings
Diseases
Aspergillus spp. (storage mold), 123
Botryotinia fuckeliana (gray mold), 125
Botrytis spp. (storage mold), 123
Cylindrocladium spp. (damping-off fungus), 121
Fusarium spp. (damping-off fungus, storage mold), 121, 123
Phytophthora spp. (damping-off fungus), 121
Pythium spp. (damping-off fungus), 121
Rhizoctonia solani (damping-off fungus), 121

***Castilleja* spp. (Indian paint-brush)**
Cronartium coleosporioides (leaf rust, stalactiform blister rust of pines), 95

***Cerastium* spp. (chickweed)**
Melampsorella caryophyllacearum (leaf rust, witches' broom rust of fir), 53

***Chamaedaphne calyculata* (L.) Moench (leatherleaf)**
Chrysomyxa cassandrae (leaf rust, spruce needle rust), 43, 45

***Comandra umbellata* (L.) Nutt. var. *pallida* (A. DC.) M.E. Jones (bastard toad-flax)**
Cronartium comandrae (leaf rust, comandra blister rust of pines), 97

***Comptonia peregrina* (L.) Coult. (sweet fern)**
Cronartium comptoniae (leaf rust, sweet fern blister rust of pines), 99

Conifers (Softwood trees)
Foliage and buds
Insects

Acleris variana (eastern blackheaded budworm), 9
Chionaspis pinifoliae (pine needle scale), 27
Dioryctria abietivorella (fir coneworm), 129
Dioryctria reniculelloides (spruce coneworm), 129
Lymantria dispar (gypsy moth), 143
Oligonychus ununguis (spruce spider mite), 25
Roots, stems, and branches
Insects
Camponotus herculeanus (red and black carpenter ant), 69
Cinara spp. (aphids), 79
Hylobius pinicola (Couper collar weevil), 63
Hylobius warreni (Warren root collar weevil), 63
Monochamus scutellatus (whitespotted sawyer beetle), 65
Sirex spp. (horntails), 69
Urocerus spp. (horntails), 69
Xeris spp. (horntails), 69
Diseases
Anisomyces odoratus (brown cubical pocket rot), 110
Armillaria calvescens (Armillaria root rot), 115
Armillaria mellea (Armillaria root rot), 115
Armillaria ostoyae (Armillaria root rot), 109, 110, 115
Armillaria sinapina (Armillaria root rot), 115
Coniophora puteana (brown cubical rot), 109, 110, 111
Echinodontium tinctorium (Indian paint fungus), 109, 110, 111
Fomitopsis officinalis (quinine conk), 109, 110
Fomitopsis pinicola (red belt fungus), 109, 110
Gloeophyllum sepiarium (slash conk), 109, 110
Gremmeniella abietina (Scleroderris canker), 105
Haematostereum sanguinolentum (bleeding fungus), 109, 110
Hirschioporus abietinus (purple conk), 109, 110
Inonotus circinatus (Tomentosus root and butt rot), 109, 110, 113
Inonotus tomentosus (Tomentosus root and butt rot), 109, 110, 113
Peniophora pseudopini (pink stain), 110
Phaeolus schweinitzii (velvet top fungus), 109, 111
Phellinus pini (red ring rot), 109, 111
Pholiota alnicola (white rot), 109, 111
Serpula himantioides (brown cubical rot), 109, 111
Sphaeropsis sapinea (Sphaeropsis blight), 103
Seeds and cones
Dioryctria abietivorella (fir coneworm), 129
Dioryctria reniculelloides (spruce coneworm), 129
Seedlings
Insects
Nomophila nearctica (celery stalkworm), 119
Otiorhynchus ovatus (strawberry root weevil), 119
Tipulidae (crane flies), 119
Diseases
Aspergillus spp. (storage mold), 123
Botryotinia fuckeliana (gray mold), 125
Botrytis spp. (storage mold), 123
Cylindrocarpon spp. (storage mold), 123
Cylindrocladium spp. (damping-off fungus), 121
Epicoccum spp. (storage mold), 123
Fusarium spp. (damping-off fungus, storage mold), 121, 123
Herpotrichia juniperi (brown felt blight), 31
Penicillium spp. (storage mold), 123
Phytophthora spp. (damping-off fungus), 121
Pythium spp. (damping-off fungus), 121
Rhizoctonia spp. (storage mold), 123
Rhizoctonia solani (damping-off fungus), 121
Rhizopus spp. (storage mold), 123

***Cotoneaster* spp. (cotoneaster, introduced)**
Foliage and buds
see Broadleaf trees
Roots, stems, and branches
Insects

see Broadleaf trees
Diseases
Erwinia amylovora (fire blight), 219
see also Broadleaf trees

Crataegus spp. (hawthorn)
Foliage and buds
Insects
Eriosoma lanigerum (woolly apple aphid), 169
see also Broadleaf trees
Diseases
see Broadleaf trees
Roots, stems, and branches
Insects
see Broadleaf trees
Diseases
Erwinia amylovora (fire blight), 219
see also Broadleaf trees

***Epilobium angustifolium* L. (fireweed)**
Pucciniastrum epilobii (fir needle rust), 53

***Fraxinus* spp. (ash)**
***pennsylvanica* Marsh. var. *subintegerrima* (Vahl) Fern. (green ash)**
Foliage and buds
Insects
Aceria fraxiniflora (ash flower gall mite), 171
Contarinia canadensis (ash midrib gall midge), 169
Tropidosteptes amoenus (ash plant bug), 167
see also Broadleaf trees
Diseases
see Broadleaf trees
Roots, stems, and branches
Insects
Hylesinus aculeatus (eastern ash bark beetle), 211
Hylesinus californicus (western ash bark beetle), 211
Hylesinus criddlei (Criddle's bark beetle), 211
Podosesia syringae (ash borer), 205
see also Broadleaf trees
Diseases
see Broadleaf trees
Seedlings
see Broadleaf trees

***Geocaulon lividum* (Richards.) Fern. (northern bastard toad-flax)**
Cronartium comandrae (leaf rust, comandra blister rust of pines), 97
***Gymnocarpium dryopteris* (L.) Newm. (oak fern)**
Hyalopsora aspidiotus (fir needle rust), 53
Uredinopsis phegopteridis (fir needle rust), 53

***Juniperus* spp. (juniper)**
***communis* L. var. *depressa* Pursh (common juniper)**
***horizontalis* Moench (creeping juniper)**
***scopulorum* Sarg. (Rocky Mountain juniper)**
Foliage and buds
Insects
see Conifers
Roots, stems, and branches
Diseases
Gymnosporangium clavariiforme (stem rust), 187, 189
Gymnosporangium clavipes (stem rust), 187, 189
Gymnosporangium nelsonii (stem gall rust), 187, 189
Gymnosporangium nidus-avis (needle rust, causes witches' brooms), 187, 189
Herpotrichia juniperi (brown felt blight), 31

***Larix* spp. (larch)**
***laricina* (Du Roi) K. Koch (tamarack)**
***lyallii* Parl. (alpine larch)**
Foliage and buds
Insects
Coleophora laricella (larch casebearer), 23
Pristiphora erichsonii (larch sawfly), 11
see also Conifers
Diseases
Melampsora albertensis (needle rust, poplar leaf rust), 183
Melampsora occidentalis (needle rust, poplar leaf rust), 183
Roots, stems, and branches
Insects
Adelges lariciatus (spruce gall adelgid), 81

Adelges strobilobius (pale spruce gall adelgid), 81
Dendroctonus simplex (eastern larch beetle), 61
see also Conifers
Diseases
Arceuthobium pusillum (eastern dwarf mistletoe), 87
see also Conifers
Seeds and cones
Insects
Megastigmus spp. (seed chalcids), 131
Strobilomyia laricis (cone maggot), 131
Strobilomyia viaria (cone maggot), 131
see also Conifers

***Ledum* spp. (Labrador tea)**
Chrysomyxa ledicola (leaf rust, spruce needle rust), 43, 45
Chrysomyxa nagodhii (leaf rust, spruce needle rust), 43, 45
Chrysomyxa neoglandulosi (leaf rust, spruce needle rust), 43, 45
Chrysomyxa woroninii (witches' broom rust, spruce shoot rust), 43, 45

***Lysimachia* spp. (loosestrife)**
Mordvilkoja vagabunda (poplar vagabond aphid), 173

***Malus* spp. (apple, crab apple)**
***pumila* (L.) Mill. (apple)**
Foliage and buds
Insects
Eriosoma lanigerum (woolly apple aphid), 169
see also Broadleaf trees
Diseases
Phyllactinia guttata (powdery mildew), 185
see also Broadleaf trees
Roots, stems, and branches
Insects
see Broadleaf trees
Diseases
Erwinia amylovora (fire blight), 219
see also Broadleaf trees

***Melampyrum lineare* Desr. (cow-wheat)**
Cronartium coleosporioides (leaf rust, stalactiform blister rust of pines), 95

***Moneses uniflora* (L.) A. Gray (one-flowered wintergreen)**
Chrysomyxa pirolata (leaf rust, spruce cone rust), 133

***Myrica gale* L. (sweet gale)**
Cronartium comptoniae (leaf rust, sweet fern blister rust of pines), 99

***Orthilia secunda* (L.) House (one-sided wintergreen)**
Chrysomyxa pirolata (leaf rust, spruce cone rust), 133

***Orthocarpus luteus* Nutt. (yellow owl-clover)**
Cronartium coleosporioides (leaf rust, stalactiform blister rust of pines), 95

***Pedicularis bracteosa* Benth. (lousewort)**
Cronartium coleosporioides (leaf rust, stalactiform blister rust of pines), 95

***Picea* spp. (spruce)**
***engelmannii* Parry (Engelmann spruce)**
***glauca* (Moench) Voss (white spruce)**
***mariana* (Mill.) B.S.P. (black spruce)**
***pungens* Engelm. (blue spruce, introduced)**
***rubens* Sarg. (red spruce)**
Foliage and buds
Insects
Choristoneura fumiferana (spruce budworm), 5
Choristoneura pinus (jack pine budworm), 7
Gilpinia hercyniae (European spruce sawfly), 15
Neodiprion abietis (balsam fir sawfly), 15
Pikonema alaskensis (yellowheaded spruce sawfly), 13
Rhabdophaga swainei (spruce bud midge), 29
Taniva albolineana (spruce needleminer), 17
Zeiraphera canadensis (spruce budmoth), 9
see also Conifers
Diseases
Chrysomyxa arctostaphyli (yellow witches' broom), 45, 47
Chrysomyxa cassandrae (needle rust), 43, 45
Chrysomyxa ledicola (needle rust), 43, 45

Chrysomyxa nagodhii (needle rust), 43, 45
Chrysomyxa neoglandulosi (needle rust), 43, 45
Chrysomyxa weirii (needle rust), 43, 45
Chrysomyxa woroninii (shoot rust), 43, 45
Herpotrichia juniperi (brown felt blight), 31
Isthmiella crepidiformis (needle cast), 41
Lirula macrospora (needle cast), 41
Lophodermium piceae (needle cast), 41
Lophophacidium hyperboreum (snow blight), 41
Melampsora albertensis (needle rust, poplar leaf rust), 183
Melampsora occidentalis (needle rust, poplar leaf rust), 183
Pucciniastrum americanum (needle rust), 43, 45
Pucciniastrum arcticum (needle rust), 43
Pucciniastrum sparsum (spruce–bearberry rust), 43, 45
see also Conifers
Roots, stems, and branches
Insects
Adelges cooleyi (Cooley spruce gall adelgid), 81
Adelges lariciatus (spruce gall adelgid), 81
Adelges strobilobius (pale spruce gall adelgid), 81
Dendroctonus rufipennis (spruce beetle), 59
Ips pini (pine engraver), 61
Mayetiola piceae (spruce gall midge), 83
Monochamus notatus (northeastern sawyer beetle), 65
Physokermes piceae (spruce bud scale), 79
Pineus coloradensis (hard pine adelgid), 79, 119
Pineus similis (ragged spruce gall adelgid), 81
Pissodes strobi (white pine weevil), 71
Tetropium cinnamopterum (wood borer), 67
Tetropium parvulum (northern spruce borer), 67
see also Conifers
Diseases
Arceuthobium americanum (lodgepole pine dwarf mistletoe), 85
Arceuthobium pusillum (eastern dwarf mistletoe), 85, 87
Burls (cause unknown), 89
Chondrostereum purpureum (silver leaf), 193
Chrysomyxa arctostaphyli (yellow witches' broom), 45, 47
Leucostoma kunzei (Leucostoma canker), 107
see also Conifers
Seeds and cones
Insects
Cydia strobilella (spruce seed moth), 129
Megastigmus atedius (spruce seed chalcid), 131
Strobilomyia neanthracina (spruce cone maggot), 131
see also Conifers
Diseases
Chrysomyxa pirolata (spruce cone rust), 133
Seedlings
Insects
Ochropleura plecta (cutworm), 119
Phyllophaga spp. (white grubs), 119
Pineus coloradensis (hard pine adelgid), 79, 119
see also Conifers
Diseases
see Conifers

***Pinus* spp., hard pines**
***banksiana* Lamb. (jack pine)**
***contorta* Dougl. var. *latifolia* Engelm. (lodgepole pine)**
***echinata* P. Mill. (shortleaf pine)**
***jeffreyi* Grev. & Balf. (Jeffrey pine)**
***mugo* Turra var. *mughus* Zenari (mugho pine, introduced)**
***muricata* D. Don (Bishop pine)**
***nigra* Arnold (Austrian pine, Corsican pine, introduced)**
***pinaster* Ait. (maritime pine)**
***ponderosa* Laws. (ponderosa pine)**
***radiata* D. Don (Monterey pine)**
***resinosa* Ait. (red pine)**
***sylvestris* L. (Scots pine, introduced)**
Foliage and buds
Insects
Argyrotaenia tabulana (jack pine tube moth), 21
Choristoneura pinus (jack pine budworm), 7

Coleotechnites starki (northern lodgepole needleminer), 19
Diprion similis (introduced pine sawfly), 15
Neodiprion maurus (sawfly), 15
Neodiprion nanulus nanulus (red pine sawfly), 15
Neodiprion swainei (Swaine jack pine sawfly), 15
see also Conifers
Diseases
Coleosporium asterum (needle rust), 39
Davisomycella ampla (needle cast), 33, 35
Elytroderma deformans (Elytroderma needle cast), 35, 37
Hendersonia pinicola (needle fungus), 33
Leptomelanconium cinereum (needle fungus), 35
Lophodermella concolor (needle cast), 33, 35
Lophodermella montivaga (needle cast), 35
Lophodermium pinastri (needle cast), 33, 35
Melampsora albertensis (needle rust, poplar leaf rust), 183
Melampsora occidentalis (needle rust, poplar leaf rust), 183
Naemacyclus minor (needle cast), 33
Naemacyclus niveus (Naemacyclus needle cast), 33, 35
Neopeckia coulteri (brown felt blight), 31
Phaeoseptoria contortae (needle fungus), 33
Thyriopsis halepensis (needle spot), 35
see also Conifers
Roots, stems, and branches
Insects
Dendroctonus ponderosae (mountain pine beetle), 57
Eucosma gloriola (eastern pine shoot borer), 77
Hylobius radicis (pine root collar weevil), 63
Ips pini (pine engraver), 61
Petrova albicapitana (northern pitch twig moth), 75
Petrova metallica (metallic pitch blister moth), 75
Physokermes piceae (spruce bud scale), 79
Pineus coloradensis (hard pine adelgid), 79, 119
Pissodes strobi (white pine weevil), 71
Pissodes terminalis (lodgepole terminal weevil), 73
Rhyacionia granti (jack pine shoot borer), 77
Rhyacionia sonia (yellow jack pine shoot borer), 77
Tetropium cinnamopterum (wood borer), 67
Tetropium parvulum (northern spruce borer), 67
Toumeyella parvicornis (pine tortoise scale), 79
see also Conifers
Diseases
Arceuthobium americanum (lodgepole pine dwarf mistletoe), 85
Arceuthobium pusillum (eastern dwarf mistletoe), 85, 87
Atropellis piniphila (Atropellis canker), 91
Burls (cause unknown), 89
Cronartium coleosporioides (stalactiform blister rust), 95
Cronartium coleosporioides f. *album* (white-spored stalactiform blister rust), 95
Cronartium comandrae (comandra blister rust), 97
Cronartium comptoniae (sweet fern blister rust), 99
Cronartium quercuum (eastern gall rust, pine–oak gall rust), 101
Endocronartium harknessii (western gall rust), 101
Neopeckia coulteri (brown felt blight), 31
Ophiostoma clavigerum (blue stain), 57
Ophiostoma huntii (blue stain), 57
Ophiostoma minus (blue stain), 57
see also Conifers
Seeds and cones
Insects
Cydia toreuta (eastern pine seedworm), 129
see also Conifers
Seedlings
Insects
Ochropleura plecta (cutworm), 119
Phyllophaga spp. (white grubs), 119
Pineus coloradensis (hard pine adelgid), 79, 119
see also Conifers
Diseases
Neopeckia coulteri (brown felt blight), 31
see also Conifers

***Pinus* spp., soft pines**
***albicaulis* Engelm. (whitebark pine)**
***edulis* Engelm. (pinyon pine)**
***flexilis* James (limber pine)**
***monticola* Dougl. (western white pine)**
***strobus* L. (eastern white pine)**

Foliage and buds
Insects
Argyrotaenia pinatubana (pine tube moth), 21
Argyrotaenia tabulana (jack pine tube moth), 21
Diprion similis (introduced pine sawfly), 15
Neodiprion nanulus nanulus (red pine sawfly), 15
Neodiprion swainei (Swaine jack pine sawfly), 15
see also Conifers
Diseases
Bifusella linearis (needle cast), 33, 35
Elytroderma deformans (Elytroderma needle cast), 37
Lophodermium nitens (needle cast), 33, 35
Lophodermium pinastri (needle cast), 33, 35
Melampsora albertensis (needle rust, poplar leaf rust), 183
Melampsora occidentalis (needle rust, poplar leaf rust), 183
Naemacyclus niveus (Naemacyclus needle cast), 33, 35
Neopeckia coulteri (brown felt blight), 31
Roots, stems, and branches
Insects
Dendroctonus ponderosae (mountain pine beetle), 57
Eucosma gloriola (eastern pine shoot borer), 77
Hylobius radicis (pine root collar weevil), 63
Ips pini (pine engraver), 61
Monochamus notatus (northeastern sawyer beetle), 65
Pissodes strobi (white pine weevil), 71
Toumeyella parvicornis (pine tortoise scale), 79
see also Conifers
Diseases
Arceuthobium americanum (lodgepole pine dwarf mistletoe), 85
Arceuthobium pusillum (eastern dwarf mistletoe), 87
Atropellis piniphila (Atropellis canker), 91
Cronartium ribicola (white pine blister rust), 93
see also Conifers
Seeds and cones
see Conifers

***Populus* spp. (aspen, poplar)**
***angustifolia* James (narrowleaf cottonwood)**
***balsamifera* L. (balsam poplar)**
***deltoides* Marsh. (eastern cottonwood)**
***grandidentata* Michx. (largetooth aspen)**
***nigra* L. var *italica* Muenchh. (Lombardy poplar)**
***tremuloides* Michx. (aspen, trembling aspen)**
***trichocarpa* Torr. & Gray (black cottonwood)**
Foliage and buds
Insects
Aceria parapopuli (poplar bud gall mite), 171
Aceria nr. *dispar* (gall mite), 171
Altica spp. (flea beetles), 159
Chaitophorus populicola (smokeywinged poplar aphid), 167
Chrysomela crotchi (aspen leaf beetle), 159
Chrysomela falsa (leaf beetle), 159
Chrysomela scripta (cottonwood leaf beetle), 159
Enargia decolor (aspen twoleaf tier), 151, 152, 153
Gonioctena americana (American aspen beetle), 157
Mordvilkoja vagabunda (poplar vagabond aphid), 173
Nematus ventralis (sawfly), 155
Pemphigus spp. (gall aphids), 173
Pemphigus populitransversus (poplar petiole gall aphid), 173
Pemphigus populivenae (gall aphid), 173
Phratora spp. (leaf beetles), 159
Phyllocnistis populiella (aspen serpentine leafminer), 165
Phyllonorycter nr. *salicifoliella*, 165
Phyllonorycter nr. *nipigon*, 165
Pseudexentera oregonana (aspen leafroller), 151, 152, 153
Tricholochmaea decora (gray willow leaf beetle), 157
see also Broadleaf trees
Diseases
Ciborinia whetzelii (ink spot), 177, 179
Linospora tetraspora (leaf blight), 177, 179
Marssonina balsamiferae (leaf spot), 177, 179
Marssonina populi (leaf spot), 177, 179
Marssonina tremuloides (leaf spot), 177, 179
Melampsora albertensis (leaf rust), 183
Melampsora medusae (leaf rust), 183

Melampsora occidentalis (leaf rust), 183
Mycosphaerella populicola (leaf spot), 177, 179
Mycosphaerella populorum (leaf spot), 177, 179
Septogloeum rhopaloideum (leaf spot), 177, 179
Uncinula adunca (powdery mildew), 185
Venturia macularis (leaf and twig blight), 177, 179, 181
Venturia populina (leaf and twig blight), 177, 179, 181
see also Broadleaf trees
Roots, stems, and branches
Insects
Agrilus liragus (bronze poplar borer), 203
Camponotus herculeanus (red and black carpenter ant), 69
Cryptorhynchus lapathi (poplar and willow borer), 201
Saperda calcarata (poplar borer), 199
Sesia tibialis (cottonwood crown borer), 205
see also Broadleaf trees
Diseases
Diplodia tumefaciens (Diplodia gall and rough-bark), 215
Entoleuca mammata (Hypoxylon canker), 217
Peniophora polygonia (pink stain rot), 225, 228
Phellinus tremulae (white rot, false tinder conk), 225, 228
Rhytidiella moriformis (rough-bark), 215
see also Broadleaf trees
Seedlings
see Broadleaf trees

***Prunus* spp. (cherry, plum)**
***pensylvanica* L.f. (pin cherry)**
***tomentosa* Thunb. (Nanking cherry, introduced)**
***virginiana* L. (choke cherry)**
Foliage and buds
Insects
see Broadleaf trees
Diseases
Phyllactinia guttata (powdery mildew), 185
see also Broadleaf trees
Roots, stems, and branches
Insects
see Broadleaf trees
Diseases
Apiosporina morbosa (black knot of cherry), 221
see also Broadleaf trees

***Pseudotsuga menziesii* (Mirb.) Franco (Douglas-fir)**
Foliage and buds
Insects
Argyrotaenia tabulana (jack pine tube moth), 21
Neodiprion abietis (balsam fir sawfly), 15
Zeiraphera canadensis (spruce budmoth), 9
see also Conifers
Diseases
Melampsora albertensis (needle rust, poplar leaf rust), 183
Melampsora occidentalis (needle rust, poplar leaf rust), 183
Roots, stems, and branches
Insects
Adelges cooleyi (Cooley spruce gall adelgid), 81
Dendroctonus pseudotsugae (Douglas-fir beetle), 61
see also Conifers
Diseases
see Conifers
Seeds and cones
Insects
Megastigmus spp. (seed chalcids), 131
see also Conifers
Seedlings
see Conifers

***Pyrola asarifolia* Michx. (common pink wintergreen)**
Chrysomyxa pirolata (leaf rust, spruce cone rust), 133

***Pyrus* spp. (pear)**
Foliage and buds
Diseases
Phyllactinia guttata (powdery mildew), 185
see also Broadleaf trees
Roots, stems, and branches
Diseases
Erwinia amylovora (fire blight), 219
see also Broadleaf trees

***Quercus* spp. (oak)**
***macrocarpa* Michx. (bur oak)**
Foliage and buds
Insects
Archips fervidana (oak webworm), 147
Callirhytis nr. *flavipes* (gall wasp), 175
Corythucha arcuata (oak lace bug), 167
Oligocentria lignicolor (lacecapped caterpillar), 151, 152, 153
Symmerista canicosta (redhumped oakworm), 151, 152, 153
see also Broadleaf trees
Diseases
see Broadleaf trees
Roots, stems, and branches
Insects
Callirhytis nr. *flavipes* (gall wasp), 175
see also Broadleaf trees
Diseases
see Broadleaf trees
Seeds
Insects
Curculio iowensis (weevil), 231
Seedlings
see Broadleaf trees

***Rhinanthus minor* L. (yellow rattle)**
Cronartium coleosporioides (leaf rust, stalactiform blister rust of pines), 95

***Ribes* spp. (currant, gooseberry)**
Cronartium ribicola (leaf rust, white pine blister rust), 93
Puccinia caricina (leaf rust), 93

***Rubus* spp. (raspberry)**
Foliage and buds
Pucciniastrum americanum (spruce–raspberry rust), 43, 45
Roots, stems, and branches
Erwinia amylovora (fire blight), 219
see also Broadleaf trees

***Salix* spp. (willow)**
Foliage and buds
Insects
Altica spp. (flea beetles), 159
Chaitophorus populicola (smokeywinged poplar aphid), 167
Chrysomela falsa (leaf beetle), 159
Chrysomela scripta (cottonwood leaf beetle), 159
Enargia decolor (aspen twoleaf tier), 151, 152, 153
Micrurapteryx salicifoliella (willow leafminer), 165
Nematus calais (sawfly), 155
Nematus limbatus (sawfly), 155
Nematus ventralis (sawfly), 155
Phratora spp. (leaf beetles), 159
Pontania proxima (willow redgall sawfly), 175
Tricholochmaea decora (gray willow leaf beetle), 157
see also Broadleaf trees
Diseases
Ciborinia foliicola (black vein), 195
Melampsora abieti-capraearum (leaf rust, fir needle rust), 53
Phyllactinia guttata (powdery mildew), 185
Rhytisma salicinum (tar spot), 195
Uncinula adunca (powdery mildew), 185
see also Broadleaf trees
Roots, stems, and branches
Insects
Cryptorhynchus lapathi (poplar and willow borer), 201
Euura atra (smaller willowshoot sawfly), 207
Saperda calcarata (poplar borer), 199
Sesia tibialis (cottonwood crown borer), 205
see also Broadleaf trees
Diseases
see Broadleaf trees
Seedlings
see Broadleaf trees

***Solidago* spp. (goldenrod)**
Coleosporium asterum (leaf rust, pine needle rust), 39

***Sorbus* spp. (mountain ash)**
Foliage and buds
Insects
Eriosoma lanigerum (woolly apple aphid), 169
Pristiphora geniculata (mountain ash sawfly), 155
see also Broadleaf trees
Diseases
Phyllactinia guttata (powdery mildew), 185
see also Broadleaf trees
Roots, stems, and branches
Insects

see Broadleaf trees
Diseases
Erwinia amylovora (fire blight), 219
see also Broadleaf trees

***Stellaria* spp. (chickweed)**
Melampsorella caryophyllacearum (leaf rust, witches' broom rust of fir), 53

***Syringa* spp. (lilac, introduced)**
Foliage and buds
Insects
see Broadleaf trees
Diseases
Microsphaera penicillata (powdery mildew), 185
see also Broadleaf trees
Roots, stems, and branches
Insects
Podosesia syringae (ash borer), 205
see also Broadleaf trees
Diseases
see Broadleaf trees

***Tsuga heterophylla* (Raf.) Sarg. (western hemlock)**
Foliage and buds
Insects
Argyrotaenia tabulana (jack pine tube moth), 21
see also Conifers
Roots, stems, and branches
see Conifers

***Ulmus* spp. (elm)**
***americana* L. (American elm, white elm)**
***pumila* L. (Siberian elm, introduced)**
Foliage and buds
Insects
Aceria ulmi (mite), 171
Eriosoma americanum (woolly elm aphid), 169
Eriosoma lanigerum (woolly apple aphid), 169
see also Broadleaf trees
Diseases
see Broadleaf trees
Roots, stems, and branches
Insects
Hylurgopinus rufipes (native elm bark beetle), 213
Scolytus multistriatus (smaller European elm bark beetle), 213
see also Broadleaf trees
Diseases
Dothiorella ulmi (wilt disease), 213
Ophiostoma novo-ulmi (Dutch elm disease), 213
Ophiostoma ulmi (Dutch elm disease), 213
Verticillium albo-atrum (wilt disease), 213
see also Broadleaf trees

***Vaccinium* spp. (blueberry, cranberry, huckleberry, grouseberry)**
Pucciniastrum goeppertianum (stem rust, fir needle rust), 53

***Viburnum lentago* (L.) (viburnum)**
Coleosporium viburni (leaf rust, pine needle rust), 39

In the following index, chapter headings are in uppercase letters.

A

Abies spp. (fir), 9, 49, 53, 63, 65, 69, 129
balsamea (balsam fir), 5, 9, 15, 49, 53, 65, 67, 131
lasiocarpa (alpine fir, subalpine fir), 31, 49, 53, 111
Aceria nr. *dispar* (gall mite), 171
Aceria fraxiniflora (ash flower gall mite), 171
Aceria parapopuli (poplar bud gall mite), 171
Aceria ulmi (gall mite), 171
Acer spp. (maple), 185
negundo (Manitoba maple), 149, 153, 167, 169, 231
Acleris variana (eastern blackheaded budworm), 9
Actinorrhizae, 253
Adelges cooleyi (Cooley spruce gall adelgid), 81, 233
Adelges lariciatus (spruce gall adelgid), 81
Adelges strobilobius (pale spruce gall adelgid), 81
ADELGID GALLS ON SPRUCE, 80–81
Agricultural chemicals, 253
Agrilus anxius (bronze birch borer), 203
Agrilus liragus (bronze poplar borer), 203
Alces alces (moose), 259
Alder (*Alnus* spp.), 159, 185, 201, 203, 253
Alnus spp. (alder), 159, 185, 201, 203, 253
Alpine bearberry (*Arctostaphylos rubra*), 45
Alpine fir (*Abies lasiocarpa*), 31, 49, 53, 111
Alpine larch (*Larix lyallii*), 131
Alsophila pometaria (fall cankerworm), 149
Altica spp. (flea beetles), 159
ambiens, 159
Ambermarked birch leafminer (*Profenusa thomsoni*), 163
Amelanchier alnifolia (saskatoon), 169, 185, 187, 191, 219, 221
American aspen beetle *(Gonioctena americana)*, 157
American elm (*Ulmus americana*), 149, 153, 171, 209, 213
Ammonia, 255
ANIMAL DAMAGE, 258–259
Anisomyces odoratus (brown cubical pocket rot), 109, 110
APHID GALLS ON POPLARS AND ASPEN, 172–173
Aphids, 79, 167, 173
APHIDS, ADELGIDS, AND SCALES, 78–79
APHIDS, LACE BUGS, AND PLANT BUGS, 166–167
Apiosporina collinsii (black leaf and witches' broom), 191, 221
Apiosporina morbosa (black knot of cherry), 191, 221
Apple (*Malus* spp., *Malus pumila*), 153, 169, 185, 193, 219
Arceuthobium americanum (lodgepole pine dwarf mistletoe), 85, 87
Arceuthobium pusillum (eastern dwarf mistletoe), 85, 87
Archips argyrospila (fruit tree leafroller), 151, 152, 153
Archips cerasivorana (uglynest caterpillar), 147
Archips fervidana (oak webworm), 147
Archips negundana (larger boxelder leafroller), 151, 152, 153
Arctostaphylos rubra (alpine bearberry), 45
Arctostaphylos uva-ursi (kinnikinnick, common bearberry), 45, 47
Arge pectoralis (birch sawfly), 155
Argyrotaenia pinatubana (pine tube moth), 21
Argyrotaenia tabulana (jack pine tube moth), 21
Armillaria calvescens (Armillaria root rot), 115
Armillaria mellea (Armillaria root rot), 115
Armillaria obscura (*see Armillaria ostoyae*)
Armillaria ostoyae (Armillaria root rot), 109, 110, 115, 226, 227
ARMILLARIA ROOT ROT, 114–115
Armillaria root rot (*Armillaria* spp.), 109, 110, 115, 227, 237, 243, 247, 249
Armillaria sinapina (Armillaria root rot), 115
Armillariella mellea (*see Armillaria mellea*)
Artist's conk (*Ganoderma applanatum*), 225, 226, 227
Ash (*Fraxinus* spp.), 153, 167, 169, 171, 209, 211, 226
ASH BARK BEETLES, 210–211
Ash bark beetles (*Hylesinus* spp.), 211
Ash borer (*Podosesia syringae*), 205
Ash flower gall mite (*Aceria fraxiniflora*), 171
Ash midrib gall midge (*Contarinia canadensis*), 169
Ash plant bug (*Tropidosteptes amoenus*), 167
Aspen (*Populus tremuloides*), 137, 139, 141, 143, 152, 153, 157, 159, 165, 171, 173, 177, 179,

181, 183, 185, 199, 203, 209, 215, 217, 223, 225, 235, 253, 255, 261
Aspen leaf beetle (*Chrysomela crotchi*), 159
Aspen leafroller (*Pseudexentera oregonana*), 151, 152, 153
Aspen serpentine leafminer (*Phyllocnistis populiella*), 165
Aspen twoleaf tier (*Enargia decolor*), 151, 152, 153
Aspergillus spp. (storage mold), 123
Aster spp. (aster), 39
ATROPELLIS CANKER OF PINE, 90–91
Atropellis canker (*Atropellis piniphila*), 91
Atropellis piniphila (Atropellis canker), 91
Austrian pine (*Pinus nigra*), 63, 101

B

Bacteria, 89, 218
Bacterial blight (*Pseudomonas syringae*), 218
Balsam fir (*Abies balsamea*), 5, 9, 15, 49, 53, 65, 67, 131
Balsam fir sawfly (*Neodiprion abietis*), 15
Balsam fir seed chalcid (*Megastigmus specularis*), 131
Balsam poplar (*Populus balsamifera*), 159, 165, 173, 177, 179, 181, 183, 215, 217, 225, 226, 245, 259
Bastard toad-flax (*Comandra umbellata* var. *pallida*), 97
Bear (*Euarctos* sp. and *Ursus* sp.), 259
Beaver (*Castor canadensis*), 259
Betula spp. (birch), 147, 153, 155, 159, 163, 203, 226, 261
    papyrifera (white birch), 153, 161, 163, 201
Bifusella abietis (see Isthmiella abietis)
Bifusella crepidiformis (see Isthmiella crepidiformis)
Bifusella linearis (needle cast), 33, 35
Birch (*Betula* spp.), 147, 153, 155, 159, 163, 203, 226, 261
Birch leafminer (*Fenusa pusilla*), 163
BIRCH LEAFMINERS, 162–163
Birch sawfly (*Arge pectoralis*), 155
Birch skeletonizer (*Bucculatrix canadensisella*), 161
Bishop pine (*Pinus muricata*), 99, 101
Bjerkandera adusta (scorched conk), 225, 227
Black cottonwood (*Populus trichocarpa*), 181, 183, 215
BLACK KNOT OF CHERRY, 220–221
Black knot of cherry (*Apiosporina morbosa*), 191, 221
BLACK LEAF AND WITCHES' BROOM OF SASKATOON, 190–191
Black leaf and witches' broom (*Apiosporina collinsii*), 191, 221
Black spruce (*Picea mariana*), 7, 9, 29, 43, 85, 87, 113, 115, 133
Black vein (*Ciborinia foliicola*), 195
Bleeding fungus (*Haematostereum sanguinolentum*), 109, 110
Blueberry (*Vaccinium* spp.), 53
Blue spruce (*Picea pungens*), 17, 25, 27, 87, 107, 131, 133, 143
Blue stain (*Ophiostoma* spp.), 57
Botryotinia fuckeliana (gray mold), 123, 125
Botrytis spp. (storage mold), 123
    cinerea (*see Botryotinia fuckeliana*)
Boxelder aphid (*Periphyllus negundinis*), 167
Boxelder bug (*Leptocoris trivittatus*), 167
Boxelder leaf gall midge (*Contarinia negundifolia*), 169
Bristly black currant (*Ribes lacustre*), 93
BRONZE BIRCH BORER AND BRONZE POPLAR BORER, 202–203
Bronze birch borer (*Agrilus anxius*), 203
Bronze poplar borer (*Agrilus liragus*), 203
Brooks poplar (*Populus deltoides* × Russian), 215
Broom rust, 45
Brown cubical pocket rot (*Anisomyces odoratus*), 109, 110
Brown cubical rot, 110, 111
BROWN FELT BLIGHT, 30–31
BRUCE SPANWORM, 141
Bruce spanworm (*Operophtera bruceata*), 141
Bucculatrix canadensisella (birch skeletonizer), 161
Buffalo-berry (*Shepherdia* spp.), 253
BURLS OF SPRUCE AND PINE, 88–89
Bur oak (*Quercus macrocarpa*), 147, 152, 167, 175, 231
Bursaphelenchus xylophilus (pine wood nematode), 65

C

Caliroa cerasi (pear sawfly), 161
Callirhytis nr. *flavipes* (gall wasp), 175
Camponotus spp. (carpenter ants) 69

herculeanus (red and black carpenter ant), 69
CANKERWORMS, 148–149
Carpenter ants (*Camponotus* spp.), 69
Castilleja spp. (Indian paint-brush), 95
Castor canadensis (beaver), 259
Cecropia moth (*Hyalophora cecropia*), 151, 152, 153
Celery stalkworm (*Nomophila nearctica*), 119
Cement, 255
Cerastium spp. (chickweed), 53
Ceratocystis ulmi (*see Ophiostoma ulmi*)
Ceratostomella ulmi (*see Ophiostoma ulmi*)
Cerrena unicolor (white spongy mottled rot), 225, 227
Cervus canadensis nelsoni (elk), 259
Chaitophorus populicola (smokeywinged poplar aphid), 167
Chamaedaphne calyculata (leatherleaf), 45
CHEMICAL POLLUTANT DAMAGE, 254–255
Cherry (*Prunus* spp.), 147, 185, 191, 221, 223
Chickweed (*Cerastium* spp., *Stellaria* spp.), 53
Chionaspis furfura (scurfy scale), 209
Chionaspis pinifoliae (pine needle scale), 27
Chipmunk (*Eutamias* spp.), 259
Choke cherry (*Prunus virginiana*), 147, 191, 221
Chondrostereum purpureum (silver leaf), 193
Choristoneura conflictana (large aspen tortrix), 139, 141
Choristoneura fumiferana (spruce budworm), 5
Choristoneura pinus (jack pine budworm), 5, 7
Chrysomela crotchi (aspen leaf beetle), 159
Chrysomela falsa (leaf beetle), 159
Chrysomela scripta (cottonwood leaf beetle), 159
Chrysomyxa arctostaphyli (spruce broom rust), 43, 45, 47, 87
Chrysomyxa cassandrae (needle rust), 43, 45
Chrysomyxa ledi (needle rust), 43
Chrysomyxa ledicola (needle rust), 43, 45, 133
Chrysomyxa nagodhii (needle rust), 43, 45
Chrysomyxa neoglandulosi (needle rust), 43, 45
Chrysomyxa pirolata (spruce cone rust), 133
Chrysomyxa weirii (needle rust), 43, 45
Chrysomyxa woroninii (shoot rust), 43, 45
Ciborinia foliicola (black vein), 195
Ciborinia whetzelii (ink spot), 177, 179
Cimbex americana (elm sawfly), 155
Cinara spp. (aphids), 79
Circular tree mortality, 249
CLEARWING MOTHS, 204–205
Clethrionomys gapperi (red-backed vole), 259
Coleophora laricella (larch casebearer), 23
Coleosporium asterum (pine needle rust), 39
Coleosporium solidaginis (*see Coleosporium asterum*)
Coleosporium viburni (needle rust), 39
Coleotechnites starki (northern lodgepole needleminer), 19
Collybia velutipes (*see Flammulina velutipes*)
Comandra blister rust (*Cronartium comandrae*), 97
COMANDRA BLISTER RUST OF HARD PINES, 96–97
Comandra umbellata var. *pallida* (bastard toad-flax), 97
Common bearberry (*Arctostaphylos uva-ursi*), 45, 47
COMMON DEFOLIATING SAWFLIES, 154–155
Common juniper (*Juniperus communis* var. *depressa*), 187, 189
Common pink wintergreen (*Pyrola asarifolia*), 133
Comptonia peregrina (sweet fern), 99
Cone maggots (*Strobilomyia* spp.), 131
CONE MAGGOTS AND SEED WASPS, 130–131
CONEWORMS AND SEED MOTHS, 128–129
Coniophora puteana (brown cubical rot), 109, 110, 111
Contarinia canadensis (ash midrib gall midge), 169
Contarinia negundifolia (boxelder leaf gall midge), 169
Cooley spruce gall adelgid (*Adelges cooleyi*), 81, 235
Coriolus hirsutus (hairy conk), 225, 227
Coriolus versicolor (white soft spongy rot), 225, 227
Corsican pine (*Pinus nigra*), 63, 101
Corticium polygonium (*see Peniophora polygonia*)
Corythucha spp. (lace bugs), 167
arcuata (oak lace bug), 167
Cotoneaster spp. (cotoneaster), 161, 193, 219
Cottonwood crown borer (*Sesia tibialis*), 205
Cottonwood leaf beetle (*Chrysomela scripta*), 159
Couper collar weevil (*Hylobius pinicola*), 63
Cow-wheat (*Melampyrum lineare*), 95
Crab apple (*Malus* spp.), 185, 193, 219
Cranberry (*Vaccinium* spp.), 53
Crane flies (Tipulidae), 119

Crataegus spp. (hawthorn), 161, 169, 193, 219
Creeping juniper (*Juniperus horizontalis*), 187, 189
Criddle's ash bark beetle (*Hylesinus criddlei*), 211
Cronartium coleosporioides (stalactiform blister rust), 95, 97, 99
Cronartium coleosporioides f. *album* (white-spored stalactiform blister rust), 95
Cronartium comandrae (comandra blister rust), 97, 101
Cronartium comptoniae (sweet fern blister rust), 99
Cronartium filamentosum (limb rust), 95
Cronartium quercuum (pine–oak gall rust), 101
Cronartium ribicola (white pine blister rust), 93
Cryptorhynchus lapathi (poplar and willow borer), 201
Curculio iowensis (seed weevil), 231
Currant (*Ribes* spp.), 93
Cydia strobilella (spruce seed moth), 129
Cydia toreuta (eastern pine seedworm), 129
Cylindrocarpon spp. (storage mold), 123
Cylindrocladium spp. (damping-off fungus), 121
Cytospora spp. (Cytospora canker), 207, 223
    chrysosperma (Cytospora canker), 223
    kunzei (*see Leucostoma kunzei*)

D

Daedalea unicolor (*see Cerrena unicolor*)
DAMAGE CAUSED BY HERBICIDES, SOIL STERILANTS, AND OTHER AGRICULTURAL CHEMICALS, 252–253
Damping-off fungus, 121
DAMPING-OFF OF SEEDLINGS, 120–121
Datana ministra (yellownecked caterpillar), 151–153
Davisomycella ampla (needle cast), 33, 35
DECAY OF ASPEN, BALSAM POPLAR, AND OTHER BROADLEAF TREES, 225–228
DECAY OF CONIFERS, 108–111
Deer (*Odocoileus* spp.), 259
Dendroctonus ponderosae (mountain pine beetle), 57, 249
Dendroctonus pseudotsugae (Douglas-fir beetle), 61
Dendroctonus rufipennis (spruce beetle), 59
Dendroctonus simplex (eastern larch beetle), 61
Dibotryon morbosum (*see Apiosporina morbosa*)
Dimerosporium collinsii (*see Apiosporina collinsii*)
Dimethyl sulfide, 255
Dioryctria abietivorella (fir coneworm), 129
Dioryctria reniculelloides (spruce coneworm), 129
Diplodia gall (*Diplodia tumefaciens*), 215
DIPLODIA GALL AND ROUGH-BARK OF POPLAR, 214–215
Diplodia pinea (*see Sphaeropsis sapinea*)
Diplodia tumefaciens (Diplodia gall and rough-bark), 215
DIPRIONID SAWFLIES, 14–15
Diprion similis (introduced pine sawfly), 15
Dothiorella ulmi (wilt disease), 213
Douglas-fir (*Pseudotsuga menziesii*), 9, 15, 21, 27, 61, 65, 69, 81, 105, 129, 183
Douglas-fir beetle (*Dendroctonus pseudotsugae*), 61
DROUGHT DAMAGE, 246–247
DUTCH ELM DISEASE, 212–213
Dutch elm disease (*Ophiostoma* spp.), 213
Dwarf mistletoe, 47, 85, 87, 265

E

Eastern ash bark beetle (*Hylesinus aculeatus*), 211
Eastern blackheaded budworm (*Acleris variana*), 9
EASTERN DWARF MISTLETOE, 86–87
Eastern dwarf mistletoe (*Arceuthobium pusillum*), 85, 87
Eastern larch beetle (*Dendroctonus simplex*), 61
Eastern pine seedworm (*Cydia toreuta*), 129
Eastern pine shoot borer (*Eucosma gloriola*), 77
Eastern white pine (*Pinus strobus*), 31, 77, 87, 93
Echinodontium tinctorium (Indian paint fungus), 109, 110, 111
Elfvingia applanatum (*see Ganoderma applanatum*)
Elk (*Cervus canadensis nelsoni*), 259
Elm (*Ulmus* spp.), 147, 153, 155, 169, 213
Elm sawfly (*Cimbex americana*), 155
Elytroderma deformans (Elytroderma needle cast), 33, 35, 37
Elytroderma needle cast (*Elytroderma deformans*), 33, 35, 37
ELYTRODERMA NEEDLE CAST OF PINE, 36–37
Enargia decolor (aspen twoleaf tier), 151, 152, 153
Endocronartium harknessii (western gall rust), 95, 97, 101
Engelmann spruce (*Picea engelmannii*), 17, 31,

43, 59, 67, 71, 89, 109, 115, 131, 133
Entoleuca mammata (Hypoxylon canker), 217
Epicoccum spp. (storage mold), 123
Epilobium angustifolium (fireweed), 53
Erethizon dorsatum (porcupine), 259
Eriophyes spp. (gall mites), 171
Eriosoma americanum (woolly elm aphid), 169
Eriosoma lanigerum (woolly apple aphid), 169
Erwinia amylovora (fire blight), 219
Erysiphe aggregata (powdery mildew), 185
Euarctos sp. (bear), 259
Eucosma gloriola (eastern pine shoot borer), 77
European fruit lecanium (*Parthenolecanium corni*), 209
European spruce sawfly (*Gilpinia hercyniae*), 15
Eutamias spp. (chipmunk), 259
Euura atra (smaller willowshoot sawfly), 207

F

Fall cankerworm (*Alsophila pometaria*), 149
Fall webworm (*Hyphantria cunea*), 147
False tinder conk (*Phellinus tremulae*), 228
Fenusa pusilla (birch leafminer), 163
Fir (*Abies* spp.), 9, 49, 53, 63, 65, 69, 129
Fir coneworm (*Dioryctria abietivorella*), 129
FIRE BLIGHT, 218–219
Fire blight (*Erwinia amylovora*), 219
Fireweed (*Epilobium angustifolium*), 53
Flammula alnicola (*see Pholiota alnicola*)
Flammulina velutipes (white heartrot), 225, 227
Flea beetles (*Altica* spp.), 159
Fluorides, 255
Fomes fomentarius (tinder conk), 225, 226, 227
Fomes igniarius (*see Phellinus tremulae*)
Fomes officinalis (*see Fomitopsis officinalis*)
Fomes pini (*see Phellinus pini*)
Fomes pinicola (*see Fomitopsis pinicola*)
Fomes tinctorius (*see Echinodontium tinctorium*)
Fomitopsis officinalis (quinine conk), 109, 110
Fomitopsis pinicola (red belt fungus), 109, 110
FOREST TENT CATERPILLAR, 136–137
Forest tent caterpillar (*Malacosoma disstria*), 137, 147
Fraxinus spp. (ash), 153, 167, 169, 171, 209, 211, 226
    pennsylvanica var. *subintegerrima* (green ash), 205, 211
Frost damage, 235
Fruit tree leafroller (*Archips argyrospila*), 151, 152, 153
Fusarium spp. (damping-off fungus, storage mold), 121, 123

G

Gall aphids (*Pemphigus* spp.), 173
GALLED AND DEFORMED LEAVES, 168–169
Gall mites (*Eriophyes* spp.), 171
GALLS ON WILLOW AND OAK, 174–175
Gall wasp (*Callirhytis* nr. *flavipes*), 175
Ganoderma applanatum (artist's conk), 225, 226, 227
Geocaulon lividum (northern bastard toad-flax), 97
Gilpinia hercyniae (European spruce sawfly), 15
Gloeocoryneum cinereum (*see Leptomelanconium cinereum*)
Gloeophyllum sepiarium (slash conk), 109, 110
Goldenrod (*Solidago* spp.), 39
Gonioctena americana (American aspen beetle), 157
Gooseberry (*Ribes* spp.), 93
Gray mold (*Botryotinia fuckeliana*), 123, 125
GRAY MOLD OF STORED CONIFER SEEDLINGS, 124–125
Gray willow leaf beetle (*Tricholochmaea decora*), 157
GRAY WILLOW LEAF BEETLE AND AMERICAN ASPEN BEETLE, 156–157
Green ash (*Fraxinus pennsylvanica* var. *subintegerrima*), 205, 211
Gremmeniella abietina (Scleroderris canker), 105
Group tree mortality, 249
Grouseberry (*Vaccinium* spp.), 53
Gymnocarpium dryopteris (oak fern), 53
Gymnopilus spectabilis (white rot), 225, 227
Gymnosporangium clavariiforme (leaf and berry rust), 187, 189
Gymnosporangium clavipes (leaf and berry rust), 187, 189
Gymnosporangium nelsonii (leaf rust), 187, 189
Gymnosporangium nidus-avis (leaf rust, causes witches' broom), 187, 189
GYPSY MOTH, 142–143
Gypsy moth (*Lymantria dispar*), 143

H

Haematostereum sanguinolentum (bleeding fungus), 109, 110
HAIL DAMAGE, 244–245

Hairy conk (*Coriolus hirsutus*), 227
Hard pine adelgid (*Pineus coloradensis*), 79, 119
Hawthorn (*Crataegus* spp.), 161, 169, 193, 219
Heavy metals, 255
Hendersonia pinicola (needle fungus), 33
Herbicides, 253
Herpotrichia coulteri (*see Neopeckia coulteri*)
Herpotrichia juniperi (brown felt blight), 31
Herpotrichia nigra (*see Herpotrichia juniperi*)
Heterarthrus nemoratus (late birch leaf edgeminer), 163
Hirschioporus abietinus (purple conk), 109, 110
Hirschioporus pargamenus (white pocket sapwood decay), 225, 227
Honey mushroom, 110
Horntails (Siricidae, *Sirex, Urocerus, Xeris*), 69
HORNTAILS AND CARPENTER ANTS, 68–69
Huckleberry (*Vaccinium* spp.), 53
Hyalophora cecropia (cecropia moth), 151, 152, 153
Hyalopsora aspidiotus (needle rust), 53
Hydrogen sulfide, 255
Hylesinus aculeatus (eastern ash bark beetle), 211
Hylesinus californicus (western ash bark beetle), 211
Hylesinus criddlei (Criddle's bark beetle), 211
Hylobius pinicola (*Couper collar weevil*), 63
Hylobius radicis (pine root collar weevil), 63
Hylobius warreni (Warren root collar weevil), 63
Hylurgopinus rufipes (native elm bark beetle), 213
Hyphantria cunea (fall webworm), 147
Hypodermella abietis-concoloris (*see Lirula abietis-concoloris*)
Hypodermella ampla (*see Davisomycella ampla*)
Hypodermella concolor (*see Lophodermella concolor*)
Hypodermella montivaga (*see Lophodermella montivaga*)
Hypoxylon canker (*Entoleuca mammata*), 217
HYPOXYLON CANKER OF ASPEN, 216–217
Hypoxylon mammatum (*see Entoleuca mammata*)

I

Ice glaze damage, 243
Indian paint-brush (*Castilleja* spp.), 95
Indian paint fungus (*Echinodontium tinctorum*), 110
Ink spot (*Ciborinia whetzelii*), 177, 179
Inonotus circinatus (Tomentosus root and butt rot), 109, 110, 113
Inonotus tomentosus (Tomentosus root and butt rot), 109, 110, 113
Introduced pine sawfly (*Diprion similis*), 15
Ips pini (pine engraver), 61
Isthmiella abietis (needle cast), 49, 51
Isthmiella crepidiformis (needle cast), 41
Isthmiella quadrispora (needle cast), 49, 51

J

Jack pine (*Pinus banksiana*), 7, 15, 21, 37, 39, 63, 73, 75, 77, 79, 85, 87, 89, 95, 97, 99, 101, 103, 105
JACK PINE BUDWORM, 6–7
Jack pine budworm (*Choristoneura pinus*), 5, 7
Jack pine shoot borer (*Rhyacionia granti*), 77
JACK PINE TUBE MOTH, 20–21
Jack pine tube moth (*Argyrotaenia tabulana*), 21
Japanese elm (*Ulmus parvifolia*), 213
Jeffrey pine (*Pinus jeffreyi*), 37, 85
Juniper (*Juniperus* spp.), 31, 119, 187
Juniperus spp. (juniper), 31, 119, 187
  communis var. *depressa* (common juniper), 187, 189
  horizontalis (creeping juniper), 187, 189
  scopulorum (Rocky Mountain juniper), 189

K

Kinnikinnick (*Arctostaphylos uva-ursi*), 45, 47

L

Labrador tea (*Ledum* spp.), 43, 45
Lace bugs (*Corythucha* spp.), 167
Lacecapped caterpillar (*Oligocentria lignicolor*), 151, 152, 153
Larch (*Larix* spp.), 69, 105, 183
LARCH CASEBEARER, 22–23
Larch casebearer (*Coleophora laricella*), 23
LARCH SAWFLY, 10–11
Larch sawfly (*Pristiphora erichsonii*), 11
LARGE ASPEN TORTRIX, 138–139
Large aspen tortrix (*Choristoneura conflictana*), 139, 141
Larger boxelder leafroller (*Archips negundana*), 151, 152, 153
Largetooth aspen (*Populus grandidentata*), 181, 217
Larix spp. (larch), 69, 105, 183
  laricina (tamarack), 11, 23, 61, 63, 65, 81, 87, 131

lyallii (alpine larch), 131
occidentalis (western larch), 183
Late birch leaf edgeminer (*Heterarthrus nemoratus*), 163
LATE-SPRING FROST DAMAGE, 234–235
Leaf and berry rust (*Gymnosporangium* spp.), 187, 189
LEAF AND BERRY RUSTS OF SASKATOON AND OTHER ROSACEOUS HOSTS, 186–189
Leaf and twig blight (*Venturia* spp.), 181
LEAF AND TWIG BLIGHT OF ASPEN AND POPLAR, 180–181
Leaf beetles, 159
Leaf blight (*Linospora tetraspora*), 177, 179
Leaf rust, 177, 183, 187, 189
LEAF RUSTS OF ASPEN AND POPLAR, 182–183
LEAF SKELETONIZERS, 160–161
Leaf spot, 177, 179, 181
LEAF SPOT DISEASES OF ASPEN AND POPLAR, 176–179
Leatherleaf (*Chamaedaphne calyculata*), 45
Ledum glandulosum (glandular Labrador tea), 45
Ledum spp. (Labrador tea), 43, 45
Lenzites sepiaria (*see Gloeophyllum sepiarium*)
Lepidosaphes ulmi (oystershell scale), 209
Leptocoris trivittatus (boxelder bug), 167
Leptomelanconium cinereum (needle fungus), 33, 35
Lepus americanus (snowshoe hare), 259
Lettuce root aphid (*Pemphigus bursarius*), 173
Leucocytospora kunzei (*see Leucostoma kunzei*)
Leucoma salicis (satin moth), 145
Leucostoma canker (*Leucostoma kunzei*), 107
LEUCOSTOMA CANKER OF SPRUCE AND OTHER CONIFERS, 106–107
Leucostoma kunzei (Leucostoma canker), 107
Light brown crumbly rot (*Piptoporus betulinus*), 228
LIGHTNING DAMAGE—CIRCULAR TREE MORTALITY, 248–249
Lilac (*Syringa* spp.), 185, 205
Limber pine (*Pinus flexilis*), 31, 85, 93
Limb rust (*Cronartium filamentosum*), 95
Linospora tetraspora (leaf blight), 177, 179
Liquid hydrocarbons, 255
Lirula abietis-concoloris (needle cast), 49, 51
Lirula macrospora (needle cast), 41
Lochmaeus manteo (variable oak leaf caterpillar), 151, 152, 153
Lodgepole pine (*Pinus contorta* var. *latifolia*), 7, 19, 21, 31, 33, 37, 39, 57, 63, 73, 75, 79, 85, 89, 91, 95, 97, 99, 101, 105, 111, 115, 119, 121, 129, 235, 237, 239, 245, 249, 255, 259
LODGEPOLE PINE DWARF MISTLETOE, 84–85
Lodgepole pine dwarf mistletoe (*Arceuthobium americanum*), 85, 87
LODGEPOLE TERMINAL WEEVIL, 72–73
Lodgepole terminal weevil (*Pissodes terminalis*), 73, 235
Lombardy poplar (*Populus nigra* var. *italica*), 215
Loosestrife (*Lysimachia* spp.), 173
Lophodermella concolor (needle cast), 33, 35, 37
Lophodermella montivaga (needle cast), 33, 35
Lophodermium macrosporum (*see Lirula macrospora*)
Lophodermium nitens (needle cast), 33, 35
Lophodermium piceae (needle cast), 41
Lophodermium pinastri (needle cast), 33, 35
Lophomerum autumnale (needle cast), 49, 51
Lophophacidium hyperboreum (snow blight), 41
Lousewort (*Pedicularis bracteosa*), 95
Lymantria dispar (gypsy moth), 143
Lyophyllum ulmarium (white rot), 225, 228
Lysimachia spp. (loosestrife), 173

M

Macrophoma tumefaciens (*see Diplodia tumefaciens*)
Malacosoma disstria (forest tent caterpillar), 137, 147
Malus spp. (apple, crab apple), 153, 169, 185, 193
pumila (apple), 193, 219
Manitoba maple (*Acer negundo*), 145, 153, 167, 169, 231
Maple (*Acer* spp.), 185
Maritime pine (*Pinus pinaster*), 101
Marssonina balsamiferae (leaf spot), 177, 179
Marssonina populi (leaf spot), 177, 179
Marssonina tremuloides (leaf spot), 177, 179
Mayetiola piceae (spruce gall midge), 83
Meadow vole (*Microtus pennsylvanicus*), 259
Megastigmus spp. (seed chalcids), 131
atedius (spruce seed chalcid), 131
specularis (balsam fir seed chalcid), 131
Melampsora abieti-capraearum (needle rust), 53

Melampsora albertensis (leaf rust), 183
Melampsora medusae (leaf rust), 183
Melampsora occidentalis (leaf rust), 183
Melampsorella caryophyllacearum (witches' broom rust), 53
Melampyrum lineare (cow-wheat), 95
Merulius himantioides (*see Serpula himantioides*)
Metallic pitch blister moth (*Petrova metallica*), 75
Metal mining and smelting, 255
Methyl mercaptan, 255
Microsphaera penicillata (powdery mildew), 185
Microtus pennsylvanicus (meadow vole), 259
Micrurapteryx salicifoliella (willow leafminer), 165
MITE GALLS, 170–171
MLO (mycoplasma-like organism), 89
Moneses uniflora (one-flowered wintergreen), 133
Monochamus notatus (northeastern sawyer), 65
Monochamus scutellatus (whitespotted sawyer beetle), 65
Monterey pine (*Pinus radiata*), 99, 101
Moose (*Alces alces*), 259
Mordvilkoja vagabunda (poplar vagabond aphid), 173
Mountain ash (*Sorbus* spp.), 155, 161, 169, 185, 193, 219
Mountain ash sawfly (*Pristiphora geniculata*), 155
MOUNTAIN PINE BEETLE, 56–57
Mountain pine beetle (*Dendroctonus ponderosae*), 57, 59, 249
Mourningcloak butterfly (*Nymphalis antiopa*), 151, 152, 153
Mugho pine (*Pinus mugo* var. *mughus*), 95, 97, 99, 101
Mycosphaerella populicola (leaf spot), 177, 179
Mycosphaerella populorum (leaf spot), 177, 179
Myrica gale (sweet gale), 99, 253

N

Naemacyclus minor (needle cast), 33
Naemacyclus needle cast (*Naemacyclus niveus*), 33, 35
Naemacyclus niveus (Naemacyclus needle cast), 33, 35
Nanking cherry (*Prunus tomentosa*), 193
Narrowleaf cottonwood (*Populus angustifolia*), 181
Native elm bark beetle (*Hylurgopinus rufipes*), 213
Natural gas, 255
NECTRIA AND CYTOSPORA ASSOCIATED WITH CANKER OF BROADLEAF TREES, 222–223
Nectria canker (*Nectria cinnabarina*), 223
Nectria cinnabarina (Nectria canker), 223
Nectria galligena (target canker), 223
Needle cast, 33, 35, 37, 41, 49, 51, 267, 268
NEEDLE CASTS AND OTHER NEEDLE DISEASES OF FIR, 48–51
NEEDLE CASTS AND OTHER NEEDLE DISEASES OF PINE, 32–35
NEEDLE CASTS AND OTHER NEEDLE DISEASES OF SPRUCE, 40–41
NEEDLE DROOP OF RED PINE, 240–241
Needle fungus, 35, 51
Needle rust, 33, 39, 41, 43, 45, 49, 53, 133
NEEDLE RUST OF HARD PINES, 38–39
NEEDLE RUSTS OF FIR, 52–53
NEEDLE RUSTS OF SPRUCE, 42–45
Needle spot, 35
Nematus calais (sawfly), 155
Nematus limbatus (sawfly), 155
Nematus ventralis (sawfly), 155
Neodiprion abietis (balsam fir sawfly), 15
Neodiprion maurus (sawfly), 15
Neodiprion nanulus nanulus (red pine sawfly), 15
Neodiprion swainei (Swaine jack pine sawfly), 15
Neopeckia coulteri (brown felt blight), 31
Nitrogen oxides, 255
Nomophila nearctica (celery stalkworm), 119
Northeastern sawyer (*Monochamus notatus*), 65
Northern bastard toad-flax (*Geocaulon lividum*), 97
NORTHERN LODGEPOLE NEEDLEMINER, 18–19
Northern lodgepole needleminer (*Coleotechnites starki*), 19
Northern pitch twig moth (*Petrova albicapitana*), 75
NORTHERN SPRUCE BORER, 66–67
Northern spruce borer (*Tetropium parvulum*), 67
Nothophacidium abietinellum (snow blight), 49, 51
Nymphalis antiopa (mourningcloak butterfly), 151, 152, 153

O

Oak (*Quercus* spp.), 143, 153, 167, 231
Oak fern (*Gymnocarpium dryopteris*), 53
Oak lacebug (*Corythucha arcuata*), 167
Oak webworm (*Archips fervidana*), 147

Ochropleura plecta (cutworm), 119
Odocoileus spp. (deer), 259
Oil refining, 255
Oligocentria lignicolor (lacecapped caterpillar), 151, 152, 153
Oligonychus ununguis (spruce spider mite), 25
One-flowered wintergreen (*Moneses uniflora*), 133
One-sided wintergreen (*Orthilia secunda*), 133
Onnia circinata (*see Inonotus circinatus*)
Onnia tomentosa (*see Inonotus tomentosus*)
Operophtera bruceata (Bruce spanworm), 141
Ophiostoma clavigerum (blue stain), 57
Ophiostoma huntii (blue stain), 57
Ophiostoma minus (blue stain), 57
Ophiostoma novo-ulmi (Dutch elm disease), 213
Ophiostoma ulmi (Dutch elm disease), 213
Orthilia spp. (wintergreen), 133
  secunda (one-sided wintergreen), 133
Orthocarpus luteus (yellow owl-clover), 95
OTHER BARK BEETLES, 60–61
OTHER BUD-FEEDING MOTHS, 8–9
OTHER COMMON DEFOLIATING CATERPILLARS, 150–153
OTHER COMMON LEAF BEETLES, 158–159
Otiorhynchus ovatus (strawberry root weevil), 119
Oystershell scale (*Lepidosaphes ulmi*), 209
Ozone, 255

P

Paleacrita vernata (spring cankerworm), 149
Pale spruce gall adelgid (*Adelges strobilobius*), 81
PAN (peroxyacetyl nitrate), 255
Parthenolecanium corni (European fruit lecanium), 209
Particulates, 255
Pear (*Pyrus* spp.), 185, 193, 219
Pear sawfly (*Caliroa cerasi*), 161
Pedicularis bracteosa (lousewort), 95
Pemphigus spp. (aphids), 173
  bursarius (lettuce root aphid), 173
  populitransversus (poplar petiole gall aphid), 173
populivenae (gall aphid), 173
Penicillium spp. (storage mold), 123
Peniophora polygonia (white rot), 255, 228
Peniophora pseudopini (pink stain), 109, 110
Perenniporia fraxinophila (white rot), 225, 226
Peridermium harknessii (*see Endocronartium harknessii*)
Periphyllis negundinis (boxelder aphid), 167
Peromyscus maniculatus borealis (white-footed deer mouse), 259
Pesotum ulmi (*see Ophiostoma ulmi*)
Petrova albicapitana (northern pitch twig moth), 75
Petrova metallica (metallic pitch blister moth), 75
Phacidium abietis (snow blight), 49, 51
Phaeocryptopus nudus (needle fungus), 49, 51
Phaeolus schweinitzii (velvet top fungus), 109, 111
Phaeoseptoria contortae (needle fungus), 33
Phellinus pini (red ring rot), 109, 111
Phellinus tremulae (false tinder conk), 225, 228
Pholiota alnicola (white rot), 109, 111
Pholiota destruens (white rot), 225, 228
Pholiota spectabilis (*see Gymnopilus spectabilis*)
Pholiota squarrosa (white mottled root and butt rot), 225, 228
Phratora spp. (leaf beetles), 159
Phyllactinia guttata (powdery mildew), 185
Phyllocnistis populiella (aspen serpentine leafminer), 165
Phyllonorycter nr. *nipigon*, 165
Phyllonorycter nr. *salicifoliella*, 165
Phyllophaga spp. (white grubs), 119
Physokermes piceae (spruce bud scale), 79
Phytophthora spp. (damping-off fungus), 121
Phytoplasma, 89
Picea spp. (spruce), 13, 15, 41, 43, 45, 47, 61, 63, 65, 69, 79, 81, 89, 105, 113, 119, 129, 183, 193, 235
  engelmannii (Engelmann spruce), 17, 31, 43, 59, 67, 71, 89, 109, 115, 131, 133
  glauca (white spruce), 5, 9, 15, 17, 25, 27, 29, 31, 41, 43, 47, 59, 63, 67, 71, 83, 85, 87
  mariana (black spruce), 7, 9, 29, 43, 85, 87, 113, 115, 133
  pungens (blue spruce), 17, 25, 27, 87, 107, 131, 133, 143
  rubens (red spruce), 29
Pikonema alaskensis (yellowheaded spruce sawfly), 13
Pin cherry (*Prunus pensylvanica*), 161, 191, 221
Pine (*Pinus* spp.), 15, 27, 33, 37, 39, 61, 65, 69, 71, 89, 101, 105, 113, 119, 129, 183, 235, 261, 267, 269
Pine engraver (*Ips pini*), 61
Pine needle rust (*Coleosporium asterum*), 39
PINE NEEDLE SCALE, 26–27

Pine needle scale (*Chionaspis pinifoliae*), 27
Pine–oak gall rust (*Cronartium quercuum*), 101
Pine root collar weevil (*Hylobius radicis*), 63
PINE SHOOT BORERS, 76–77
Pine tortoise scale (*Toumeyella parvicornis*), 79
Pine tube moth (*Argyrotaenia pinatubana*), 21
Pineus coloradensis (hard pine adelgid), 79, 119
Pineus similis (ragged spruce gall adelgid), 81
Pine wood nematode (*Bursaphelenchus xylophilus*), 65
Pink stain (*Peniophora pseudopini*), 110
Pinus spp. (pine), 15, 27, 33, 37, 39, 61, 65, 69, 71, 89, 101, 105, 113, 119, 129, 183, 235, 261, 267, 269
- *albicaulis* (whitebark pine), 21, 85, 91, 93
- *banksiana* (jack pine), 7, 15, 21, 37, 39, 63, 73, 75, 77, 79, 85, 87, 89, 95, 97, 99, 101, 103, 105
- *contorta* var. *latifolia* (lodgepole pine), 7, 19, 21, 31, 33, 37, 39, 57, 63, 73, 75, 79, 85, 89, 91, 95, 97, 99, 101, 105, 111, 115, 119, 121, 129, 235, 237, 239, 245, 249, 255, 259
- *echinata* (shortleaf pine), 37, 95, 99
- *edulis* (pinyon pine), 37, 85
- *flexilis* (limber pine), 31, 85, 93
- *jeffreyi* (Jeffrey pine), 37, 85
- *monticola* (western white pine), 21, 65, 91
- *mugo* var. *mughus* (mugho pine), 95, 97, 99, 101
- *muricata* (Bishop pine), 99, 101
- *nigra* (Austrian pine, Corsican pine), 63, 101
- *pinaster* (maritime pine), 101
- *ponderosa* (ponderosa pine), 19, 37, 75, 85, 91, 95, 97, 99, 101
- *radiata* (Monterey pine), 99, 101
- *resinosa* (red pine), 7, 15, 63, 79, 87, 99, 103, 105, 107, 115, 119, 129, 241, 247
- *strobus* (eastern white pine), 21, 77, 87, 93
- *sylvestris* (Scots pine), 7, 15, 33, 39, 63, 79, 85, 95, 97, 99, 101, 105

Pinyon pine (*Pinus edulis*), 37
Piptoporus betulinus (light brown crumbly rot), 225, 226, 227
Pissodes spp. (bark weevils), 235
- *strobi* (white pine weevil), 71, 235
- *terminalis* (lodgepole terminal weevil), 73, 235

PITCH BLISTER MOTHS, 74–75
Pleurotus ulmarius (*see Lyophyllum ulmarium*)
Plum (*Prunus* spp.), 147, 185, 193, 221
Podosesia syringae (ash borer), 205
Pollaccia elegans (*see Venturia populina*)
Pollaccia radiosa (*see Venturia macularis*)
Polyporus abietinus (*see Hirschioporus abietinus*)
Polyporus adustus (*see Bjerkandera adusta*)
Polyporus betulinus (*see Piptoporus betulinus*)
Polyporus fomentarius (*see Fomes fomentarius*)
Polyporus hirsutus (*see Coriolus hirsutus*)
Polyporus pargamenus (*see Hirschioporus pargamenus*)
Polyporus schweinitzii (*see Phaeolus schweinitzii*)
Polyporus tomentosus var. *circinatus* (*see Inonotus circinatus*)
Polyporus tomentosus var. *tomentosus* (*see Inonotus tomentosus*)
Polyporus versicolor (*see Coriolus versicolor*)
Ponderosa pine (*Pinus ponderosa*), 19, 37, 75, 85, 91, 95, 97, 99, 101
Pontania proxima (willow redgall sawfly), 175
Poplar (*Populus* spp.), 145, 155, 159, 167, 171, 173, 177, 183, 185, 193, 199, 201, 203, 205, 215
POPLAR AND WILLOW BORER, 200–201
Poplar and willow borer (*Cryptorhynchus lapathi*), 201
POPLAR AND WILLOW LEAFMINERS, 164–165
POPLAR BORER, 198–199
Poplar borer (*Saperda calcarata*), 199, 203
Poplar bud gall mite (*Aceria parapopuli*), 171
Poplar petiole gall aphid (*Pemphigus populitransversus*), 173
Poplar vagabond aphid (*Mordvilkoja vagabunda*), 173
Populus spp. (aspen, poplar), 145, 155, 159, 167, 171, 173, 177, 183, 185, 193, 199, 201, 203, 205, 215
- *angustifolia* (narrowleaf cottonwood), 181
- *balsamifera* (balsam poplar), 159, 165, 173, 177, 179, 181, 183, 215, 217, 225, 245, 259
- *deltoides* × Russian (Brooks poplar), 215
- *grandidentata* (largetooth aspen), 181, 215
- *nigra* var. *italica* (Lombardy poplar), 215
- *tremuloides* (aspen), 137, 139, 141, 143, 153, 157, 159, 165, 171, 173, 177, 179, 181, 183, 185, 199, 203, 209, 215, 217,

223, 225, 235, 253, 255, 261
trichocarpa (black cottonwood), 181, 183, 215
Porcupine (*Erethizon dorsatum*), 259
Potash, 255
Powdery mildew, 185, 266, 268
POWDERY MILDEW OF BROADLEAF TREES, 184–185
Pristiphora erichsonii (larch sawfly), 11
Pristiphora geniculata (mountain ash sawfly), 155
Profenusa thomsoni (ambermarked birch leaf miner), 163
Proteoteras nr. *aesculana* (caterpillar), 231
Prunus spp. (cherry, plum), 147, 185, 193, 221
pensylvanica (pin cherry), 161, 221
tomentosa (Nanking cherry), 193
virginiana (choke cherry), 147, 221
Pseudexentera oregonana (aspen leafroller), 151, 152, 153
Pseudomonas syringae (bacterial blight), 219
Pseudotsuga menziesii (Douglas-fir), 9, 15, 21, 27, 61, 65, 69, 81, 105, 129, 183
Puccinia caricina (leaf rust), 93
Pucciniastrum spp., 43
americanum (needle rust), 43, 45, 133
arcticum (needle rust), 43, 45
epilobii (needle rust), 53
goeppertianum (needle rust), 53
sparsum (spruce–bearberry rust), 43, 45
Pulp and paper industry, 255
Purple conk (*Hirschioporus abietinus*), 110
Pyrola spp. (wintergreen), 133
asarifolia (common pink wintergreen), 133
Pyrus spp. (pear), 185, 193, 219
Pythium spp. (damping-off fungus), 121

Q

Quadraspidiotus perniciosus (San José scale), 209
Quercus spp. (oak), 143, 153, 231
macrocarpa (bur oak), 147, 152, 167, 175, 231
Quinine conk (*Fomitopsis officinalis*), 110

R

Radulodon americanus (white stringy heartrot), 225, 228
Radulum casearium (*see Radulodon americanus*)
Ragged spruce gall adelgid (*Pineus similis*), 81
Raspberry (*Rubus* spp.), 45, 219
Red and black carpenter ant (*Camponotus herculeanus*), 69
Red-backed vole (*Clethrionomys gapperi*), 259
Red-bellied sapsucker (*Sphyrapicus varius ruber*), 261
RED BELT, 238–239
Red belt fungus (*Fomitopsis pinicola*), 109, 110
Redhumped oakworm (*Symmerista canicosta*), 151, 152, 153
Red pine (*Pinus resinosa*), 7, 15, 63, 79, 87, 99, 103, 105, 107, 115, 119, 129, 241, 247
Red pine sawfly (*Neodiprion nanulus nanulus*), 15
Red ring rot (*Phellinus pini*), 109, 111
Red spruce (*Picea rubens*), 29
Red squirrel (*Tamiasciurus hudsonicus*), 259
Rhabdophaga swainei (spruce bud midge), 29
Rhinanthus minor (yellow rattle), 95
Rhizoctonia spp. (storage mold), 123
solani (damping-off fungus), 121
Rhizopus spp. (storage mold), 123
Rhyacionia granti (jack pine shoot borer), 77
Rhyacionia sonia (yellow jack pine shoot borer), 77
Rhytidiella moriformis (rough-bark), 215
Rhytisma salicinum (tar spot), 195
Ribes spp. (currant, gooseberry), 93
glandulosum (skunk currant), 93
hudsonianum (wild black currant), 93
lacustre (bristly black currant), 93
Rocky Mountain juniper (*Juniperus scopulorum*), 189
Root collar weevil, 237
ROOT COLLAR WEEVILS, 62–63
Rough-bark, 215
Rubus spp. (raspberry), 45, 219

S

Salix spp. (willow), 53, 143, 145, 147, 153, 155, 157, 159, 165, 167, 175, 185, 193, 195, 199, 201, 205, 207, 253
Saltwater spills, 255
San José scale (*Quadraspidiotus perniciosus*), 209
Saperda calcarata (poplar borer), 199
Sapsucker (*Sphyrapicus* spp.), 261
SAPSUCKER INJURY, 260–261
Sarcotrochila balsameae (snow blight), 49, 51
Saskatoon (*Amelanchier alnifolia*), 169, 185, 187, 189, 191, 219, 221
SATIN MOTH, 144–145
Satin moth (*Leucoma salicis*), 145
SAWYER BEETLES, 64–65

SCALE INSECTS, 208–209
SCLERODERRIS CANKER, 104–105
Scleroderris canker (*Gremmeniella abietina*), 105
Scleroderris lagerbergii (*see Gremmeniella abietina*)
Scolytus multistriatus (smaller European elm bark beetle), 213
Scorched conk (*Bjerkandera adusta*), 225, 227
Scots pine (*Pinus sylvestris*), 7, 15, 33, 39, 63, 79, 85, 95, 97, 99, 101, 105
Scurfy scale (*Chionaspis furfura*), 209
Seed chalcids (*Megastigmus* spp.), 131
SEED INSECTS, 230–231
Septogloeum rhopaloideum (leaf spot), 177, 179
Septoria populicola (*see Mycosphaerella populicola*)
Serpula himantioides (brown cubical rot), 109, 111
Sesia tibialis (cottonwood crown borer), 205
Shepherdia spp. (buffalo-berry), 253
Shoot rust (*Chrysomyxa woroninii*), 45
Shortleaf pine (*Pinus echinata*), 37, 95, 99
Shrew (*Sorex* spp.), 259
Siberian elm (*Ulmus pumila*), 149, 213
SILVER LEAF, 192–193
Silver leaf (*Chondrostereum purpureum*), 193
Sirex spp. (horntails), 69
Siricidae (*Sirex, Urocerus, Xeris*) (horntails, woodwasps), 69
Skunk currant (*Ribes glandulosum*), 93
Slash conk (*Gloeophyllum sepiarium*), 110
Smaller European elm bark beetle (*Scolytus multistriatus*), 213
SMALLER WILLOWSHOOT SAWFLY, 206–207
Smaller willowshoot sawfly (*Euura atra*), 207
Smokeywinged poplar aphid (*Chaitophorus populicola*), 167
Snow blight, 41, 51
Snow damage, 243
Snowshoe hare (*Lepus americanus*), 259
Sodium hydroxide, 255
Soil sterilants, 253
Solidago spp. (goldenrod), 39
Sorbus spp. (mountain ash), 155, 161, 185, 193, 219
Sorex spp. (shrew), 259
Sphaeropsis blight (*Sphaeropsis sapinea*), 103
SPHAEROPSIS (DIPLODIA) BLIGHT OF PINE AND OTHER CONIFERS, 102–103
Sphaeropsis sapinea (Sphaeropsis blight), 103
Sphyrapicus varius varius (yellow-bellied sapsucker), 261
Sphyrapicus varius ruber (red-bellied sapsucker), 261
Spring cankerworm (*Paleacrita vernata*), 149
Spruce (*Picea* spp.), 13, 15, 41, 43, 45, 47, 61, 63, 65, 69, 79, 81, 89, 105, 113, 119, 129, 183, 193, 235
Spruce–bearberry rust (*Pucciniastrum sparsum*), 45
SPRUCE BEETLE, 58–59
Spruce beetle (*Dendroctonus rufipennis*), 59
Spruce broom rust (*Chrysomyxa arctostaphyli*), 43, 45, 47, 87
SPRUCE BUD MIDGE, 28–29
Spruce bud midge (*Rhabdophaga swainei*), 29
Spruce bud scale (*Physokermes piceae*), 79
Spruce budmoth (*Zeiraphera canadensis*), 9
SPRUCE BUDWORM, 4–5
Spruce budworm (*Choristoneura fumiferana*), 5, 9
Spruce cone maggot (*Strobilomyia neanthracina*), 131
SPRUCE CONE RUST, 132–133
Spruce cone rust (*Chrysomyxa pirolata*), 133
Spruce coneworm (*Dioryctria reniculelloides*), 129
Spruce gall adelgid (*Adelges lariciatus*), 81
SPRUCE GALL MIDGE, 82–83
Spruce gall midge (*Mayetiola piceae*), 83
SPRUCE NEEDLEMINER, 16–17
Spruce needleminer (*Endothenia albolineana*), 17
Spruce needle rust, 41, 43, 45, 133
Spruce seed chalcid (*Megastigmus atedius*), 131
Spruce seed moth (*Cydia strobilella*), 129
SPRUCE SPIDER MITE, 24–25
Spruce spider mite (*Oligonychus ununguis*), 25
Stalactiform blister rust (*Cronartium coleosporioides*), 95, 97, 99
STALACTIFORM BLISTER RUST OF HARD PINES, 94–95
Stationary combustion engines, 255
Stegopezizella balsameae (*see Sarcotrochila balsameae*)
Stellaria spp. (chickweed), 53
Stereum purpureum (*see Chondrostereum purpureum*)
Stereum sanguinolentum (*see Haematostereum sanguinolentum*)
Storage mold, 123
STORAGE MOLDS OF CONIFER SEEDLINGS, 122–123
Strawberry root weevil (*Otiorhynchus ovatus*), 119

Strobilomyia laricis (cone maggot), 131
Strobilomyia neanthracina (spruce cone maggot), 131
Strobilomyia viaria (cone maggot), 131
Subalpine fir (*Abies lasiocarpa*), 31, 49, 53, 111
Sulfur dioxide, 255
Swaine jack pine sawfly (*Neodiprion swainei*), 15
Sweet fern (*Comptonia peregrina*), 99
Sweet fern blister rust (*Cronartium comptoniae*), 99
SWEET FERN BLISTER RUST OF HARD PINES, 98–99
Sweet gale (*Myrica gale*), 253
Symmerista canicosta (redhumped oakworm), 151–153
Syringa spp. (lilac), 205
  vulgaris (lilac), 185

T

Taniva albolineana (spruce needleminer), 17
Tamarack (*Larix laricina*), 11, 23, 61, 63, 65, 81, 87, 131
Tamiasciurus hudsonicus (red squirrel), 259
Target canker (*Nectria galligena*), 223
Tar spot (*Rhytisma salicinum*), 195
TAR SPOT AND BLACK VEIN OF WILLOW, 194–195
Tetropium cinnamopterum (wood borer), 67
Tetropium parvulum (northern spruce borer), 67
Thuja occidentalis (white cedar), 119
Thyriopsis halepensis (needle spot), 33, 35
Tinder conk (*Fomes fomentarius*), 227
Tipulidae (crane flies), 119
Tomentosus root and butt rot (*Inonotus* spp.), 109, 110, 113
TOMENTOSUS ROOT ROT OF CONIFERS, 112–113
Toumeyella parvicornis (pine tortoise scale), 79
Trametes odorata (*see Anisomyces odoratus*)
Trembling aspen (*Populus tremuloides*), 137, 139, 141, 143, 153, 157, 159, 165, 171, 173, 177, 179, 181, 183, 185, 199, 203, 209, 215, 217, 223, 225, 235, 253, 255, 261
Tricholochmaea decora (gray willow leaf beetle), 157
Tropidosteptes amoenus (ash plant bug), 167
Tsuga heterophylla (western hemlock), 21
Tubercularia vulgaris (*see Nectria cinnabarina*)

U

Uglynest caterpillar (*Archips cerasivorana*), 147
UGLYNEST CATERPILLAR AND FALL WEBWORM, 146–147
Ulmus spp. (elm), 147, 153, 155, 169, 213
  americana (American elm, white elm), 149, 153, 171, 209, 213
  parvifolia (Japanese elm), 213
  pumila (Siberian elm), 149, 213
Uncinula adunca (powdery mildew), 185
Uncinula salicis (*see Uncinula adunca*)
Uredinopsis phegopteridis (needle rust), 53
Urocerus spp. (horntails), 69
Ursus sp. (bear), 259

V

Vaccinium spp. (blueberry, cranberry, huckleberry, grouseberry), 53
Valsa kunzei (*see Leucostoma kunzei*)
Valsa sordida (*see Cytospora chrysosperma*)
Variable oak leaf caterpillar (*Lochmaeus manteo*), 151, 152, 153
Velvet top fungus (*Phaeolus schweinitzii*), 109, 111
Venturia macularis (leaf and twig blight), 177, 179, 181
Venturia populina (leaf and twig blight), 177, 179, 181
Venturia tremulae (*see Venturia macularis*)
Verticillium albo-atrum (wilt disease), 213
Viburnum lentago (viburnum), 39
Viburnum (*Viburnum lentago*), 39
Virus, 89

W

Warren root collar weevil (*Hylobius warreni*), 63
Western ash bark beetle (*Hylesinus californicus*), 211
Western gall rust (*Endocronartium harknessii*), 89, 95, 97, 101
WESTERN GALL RUST OF HARD PINES, 100–101
Western hemlock (*Tsuga heterophylla*), 21
Western larch (*Larix occidentalis*), 183
Western white pine (*Pinus monticola*), 21, 65, 91
Whitebark pine (*Pinus albicaulis*), 21, 85, 91, 93
White birch (*Betula papyrifera*), 161
White cedar (*Thuja occidentalis*), 119
White elm (*Ulmus americana*), 149, 153, 171, 209, 213
White-footed deer mouse (*Peromyscus maniculatus borealis*), 259
White grubs (*Phyllophaga* spp.), 119
White heartrot (*Flammulina velutipes*), 227
White mottled root and butt rot (*Pholiota squarrosa*), 225, 228

WHITE PINE BLISTER RUST, 92–93
White pine blister rust (*Cronartium ribicola*), 93
WHITE PINE WEEVIL, 70–71
White pine weevil (*Pissodes strobi*), 71, 235
White pocket sapwood decay (*Hirschioporus pargamenus*), 225, 227
White rot, 109, 111, 217, 225, 227, 228
White soft spongy rot (*Coriolus versicolor*), 225, 227
White spongy mottled rot (*Cerrena unicolor*), 225, 227
White-spored stalactiform blister rust (*Cronartium coleosporioides* f. *album*), 95
Whitespotted sawyer beetle (*Monochamus scutellatus*), 65
White spruce (*Picea glauca*), 5, 9, 15, 17, 25, 27, 29, 31, 41, 43, 47, 59, 63, 67, 71, 83, 85, 87
White stringy heartrot (*Radulodon americanus*), 225, 228
Wild black currant (*Ribes hudsonianum*), 93
Willow leafminer (*Micrurapteryx salicifoliella*), 165
Willow redgall sawfly (*Pontania proxima*), 175
Willow (*Salix* spp.), 53, 143, 145, 147, 153, 155, 157, 159, 165, 167, 175, 185, 193, 195, 199, 201, 205, 207, 253
Wilt disease, 213
Wind damage, 243
WIND, SNOW, AND ICE GLAZE DAMAGE, 242–243
WINTER DESICCATION DAMAGE, 236–237
Wintergreen, 133
Witches' broom of saskatoon (*Apiosporina collinsii*), 191, 221
Witches' broom rust (*Melampsorella caryophyllacearum*), 53
Woolly apple aphid (*Eriosoma lanigerum*), 169
Woolly elm aphid (*Eriosoma americanum*), 169

X

Xeris spp. (horntails), 69

Y

Yellow-bellied sapsucker (*Sphyrapicus varius varius*), 261
YELLOWHEADED SPRUCE SAWFLY, 12–13
Yellowheaded spruce sawfly (*Pikonema alaskensis*), 13
Yellow jack pine shoot borer (*Rhyacionia sonia*), 77
Yellownecked caterpillar (*Datana ministra*), 151, 152, 153
Yellow owl-clover (*Orthocarpus luteus*), 95
Yellow rattle (*Rhinanthus minor*), 95
YELLOW WITCHES' BROOM OF SPRUCE, 46–47
Yellow witches' broom of spruce (*Chrysomyxa arctostaphyli*), 41, 43, 45, 47, 53, 87

Z

Zeiraphera canadensis (spruce budmoth), 9